GRAPH THEORY AND ITS APPLICATIONS

Publication No. 24
of the Mathematics Research Center
United States Army
The University of Wisconsin

ERRATA

GRAPH THEORY AND ITS APPLICATIONS

Edited by Bernard Harris

page vi lines 8-13:

On the Foundations of Combinatorial Theory 167
Gian-Carlo Rota
Massachusetts Institute of Technology,
Cambridge, Massachusetts
Ronald Mullin
Florida Atlantic University, Boca Raton, Florida

should read:

On the Foundations of Combinatorial Theory: III. Theory of
Binomial Enumeration 167
Ronald Mullin
Florida Atlantic University, Boca Raton, Florida
Gian-Carlo Rota
Massachusetts Institute of Technology,
Cambridge, Massachusetts

page 167 lines 1-3:

On the Foundations of
Combinatorial Theory
GIAN-CARLO ROTA AND RONALD MULLIN

should read:

On the Foundations of Combinatorial Theory:
III. Theory of Binomial Enumeration
RONALD MULLIN AND GIAN-CARLO ROTA

Graph Theory and Its Applications

Edited by Bernard Harris

Proceedings of an Advanced Seminar
Conducted by the Mathematics Research Center,
United States Army, at the University
of Wisconsin, Madison
October 13-15, 1969

Academic Press
New York · London 1970

ACADEMIC PRESS, INC.
111 Fifth Avenue, New York, New York 10003

United Kingdom Edition published by
ACADEMIC PRESS, INC. (LONDON) LTD.
Berkeley Square House, London W1X 6BA

LIBRARY OF CONGRESS CATALOG CARD NUMBER: 71-117086

PRINTED IN THE UNITED STATES OF AMERICA

Contents

PREFACE vii

Graph Theory as a Structural Model in the
Social Sciences 1
Frank Harary
University of Michigan, Ann Arbor, Michigan

The Mystery of the Heawood Conjecture 17
J. W. T. Youngs
University of California, Santa Cruz, California

Graph Theory Algorithms 51
Ronald C. Read
University of the West Indies, Jamaica, West Indies

On Eigenvalues and Colorings of Graphs 79
Alan J. Hoffman
Thomas J. Watson Research Center,
I.B.M. Corporation, Yorktown Heights, New York

Blocking Polyhedra 93
D. R. Fulkerson
The Rand Corporation, Santa Monica, California

Connectivity in Matroids 113
W. T. Tutte
University of Waterloo, Ontario, Canada

The Use of Circuit Codes in Analog-to-Digital
Conversion 121
Victor Klee
University of Washington, Seattle, Washington

CONTENTS

Some Mapping Problems for Tournaments 133
J. W. Moon
University of Alberta, Edmonton, Canada

On Some Connections between Graph Theory and Experimental Designs and Some Recent Existence Results 149
D. K. Ray Chaudhuri
Ohio State University, Columbus, Ohio

On the Foundations of Combinatorial Theory 167
Gian-Carlo Rota
Massachusetts Institute of Technology, Cambridge, Massachusetts
Ronald Mullin
Florida Atlantic Unitversity, Boca Raton, Florida

A Composition Theorem for the Enumeration of Certain Subsets of the Symmetric Semigroup 215
Bernard Harris
University of Wisconsin, Madison, Wisconsin
Lowell Schoenfeld
SUNY at Buffalo, Amherst, New York

Symbols for the Harris-Schoenfeld Paper 253

INDEX 255

Preface

In selecting topics for our annual advanced seminar, we look for a branch of mathematics in which there has been a substantial amount of significant recent research activity. The proliferation of activity in combinatorics made us aware that this is an area eminently suitable for an advanced seminar. Thus, we felt that combinatorics in general and graph theory in particular met our requirements.

To substantiate these remarks I would like to quote from the Gibbs' lecture of Professor R. L. Wilder, published in the Bulletin of the American Mathematical Society, **75** (1969), 891-906.

"I can recall that while teaching a course in graph theory 30 or 35 years ago, I recognized that far from being the dead field that it was regarded to be at the time, it offered great potential for research by a student without a great deal of background in classical mathematics. However, the recent resurgence of research in graph theory was apparently not due to any such consideration, but to the discovery that it had applications to problems in both the natural and social sciences—a fact that I suspected Cayley knew but had no time to pursue beyond some elementary work in chemical bonds."

Therefore, we held the Advanced Seminar on Graph Theory and Its Applications on October 13-15, 1969. Eleven papers were presented at the advanced seminar and ten of these are in this volume. An additional paper by the editor and Lowell Schoenfeld was added after discussion with the publisher and the director of the Mathematics Research Center.

We hope that this volume will provide an appropriate survey of the "state of the art" in graph theory and motivate broader application as well as advancing mathematical research in the subject.

I am certain that the attendees at this advanced seminar will agree with me that this was one of the most colorful events in recent mathematical history.

There are a number of people who do not appear as authors in this volume to whom acknowledgment is certainly due. The chairmen of the various sessions included W. G. Brown, N. S. Mendelsohn, Paul Camion, and R. H. Bruck. T. C. Hu, R. A. Brualdi, and E. F. Moore assisted the editor in planning the program. A. C. Tucker provided some editorial assistance and mathematical advice to the editor. Mrs. Gladys Moran gave her customary competent performance as secretary of the program committee and, in addition, handled all the necessary physical arrangements with remarkable skill. Mrs. Dorothy Bowar is to be particularly commended for her painstaking efforts in the typing of the manuscripts and preparation of the figures for publication.

Graph Theory as a Structural Model in the Social Sciences

FRANK HARARY

Dedicated to Raymond L. Wilder on his 73rd Birthday

In his Gibbs Lecture [36], Raymond L. Wilder expressed the case for research in graph theory and its applications so eloquently that we take the liberty of quoting from it:

"Mathematics, like music, profits by cultivation in the very young, and anyone who uncovers a previously undeveloped field of mathematics that permits the student of little background to do research in it, is probably providing as great a service to mathematics as one whose research has obvious significance for either applications or for the main lines along which mathematics is being developed at the time.

After all, the value of research is also a relative thing. I can recall that while teaching a course in graph theory 30 or 35 years ago, I recognized that far from being the dead field that it was regarded to be at the time, it offered great potential for research by a student without a great deal of background in classical mathematics. However, the recent resurgence of research in graph theory was apparently not due to any such consideration, but to the discovery that it had applications to problems in both the natural and social sciences--a fact that I suspect Cayley knew but had no time to pursue beyond some elementary work on chemical bonds."

The meaning of the term "mathematical model" to a social scientist is, in general, that branch of mathematics which is suitable as an abstraction of his field of interest. Sometimes the appropriate kind of mathematics already exists and sometimes it is necessary to modify established areas of mathematics by the addition of new concepts, to make the resulting theory suitable as the abstraction of the particular field of social science at hand. In this way, the theory of graphs, sometimes intact and at other times with modifications, is being used more and more frequently as a mathematical model in various social sciences. There are two different ways in which it is so used. It may be employed, especially at first, only as a translation of real world phenomena into mathematical terms: this we call a level one model. More far reaching are those mathematical models which go beyond what amounts to setting up an axiom system which is appropriate, and seek to apply to the real world situation the theorems which are derived from the axiom systems. Optimally, from the viewpoint of a mathematician, the best mathematical models are those which will lead to new theorems, some of which may even be publishable in the mathematical journals. In this case we speak of a level two model. Both of these levels are useful in their own ways. Level one models serve at least to clarify the ideas in the subject at hand, whether it be a fragment of psychology, sociology, economics, anthropology, linguistics, geography, etc. Another way in which a level one model may be useful is that it can provide a framework for the collection of relevant data that can lead to a meaningful statistical analysis. Of course a level two model is the more far reaching as the theorems, when applied to the real world, constitute predictions of phenomena that may not even have occurred to the social scientist previously. To borrow from the physical sciences, probably the most vivid illustration of a level two mathematical model occurred in Einstein's Theory of Relativity. We will see examples of models at both levels in this review article, which we restrict to only a few social scientific topics.

We begin with some level one models in the field of geography. These include transportation networks and graphs depicting frequency of communication between cities. We then turn to a discussion of the structure of interpersonal groups and of organizations in terms of graph theory, in which the principal stress is on algorithms. The study of the sociological field of role theory can be made much more precise and definite by means of a graphical model which we describe. This involves quite a comprehensive mixture of structures including graphs, directed graphs, bipartite graphs, and networks. Because the examples of Markov chains are so numerous we indicate only two of them, namely models for consumer behavior relative to brand loyalty and brand switching, and for learning theory. We conclude with a presentation of our theory of structural balance from social psychology, which in disguise basically involves problems in the colorings of graphs, and has proved to be a level two model, and brief mention of network flows.

In general, we will use the book [18] for terms from graph theory, [24] for directed graphs, and recommend the bibliography complied by J. Turner in [19] for further sources of papers in both graph theory and its applications to social science.

Geography

We will discuss briefly four papers from the literature of geography, in which "nodal regions" is a technical term. In their innovative paper in this field Nystuen and Dacey [33] write

"The purpose of this paper is to describe a procedure for ordering and grouping cities by the magnitude and direction of the flows of goods, people, and communications between them. Pertinent geographic and graph theoretic concepts are discussed and are then used as a basis for deriving the method of isolating nodal regions."

An indication of the way in which they interpret graph theory is given by them in the quotation,

"The present problem is to develop a method capable of quantifying the degree of association between city pairs in a manner that allows identification of the networks of strongest association. These associations may be in terms of interactions that occur directly between two cities, or indirectly through one or more intermediary cities. The magnitude of the combined direct and indirect associations is measured by an index that is related to certain concepts of graph theory."

The variety of other potential applications of graph theory to geography is then indicated by them as follows:

"In the present illustration, cities are conceptualized as punctiform elements in a telephone network. Other suitable areas of application include the flow of information or material products between business firms in a metropolitan area, the flow of mail or freight between cities in a region, the interpersonal relations between the inhabitants of a city or the political structure that connects federal, state and local governments."

In this particular study, a digraph is taken to have its points represent cities and its arcs (directed lines) stand for the largest flow (of telephone calls) from a subordinate city to a superordinate city. The result is a digraph in which every weak component is a tree to a point.

There is a folk story, which is apparently untrue, to the effect that mapmakers were the first to happen upon the Four Color Conjecture (see Chapter 1 of [18]). On the other hand, the article by Dacey [6] on k-color maps is a statistical study by a geographer of map coloring which extends earlier work of the statistician, P. A. Moran.

In the report by Nystuen [32], a transportation network is defined as a graph in which the points are the actual intersections of two or more roads and each line is a segment of a road between two of these points. Nystuen then interprets the degree of a point as the number of roads converging at that intersection (and makes quite plausible assumptions for correlating the quantity of traffic with the degree of a point); a connected component is illustrated by means of a road network within a rubber plantation which has no surface road to the outside world; an elementary contraction is the identification of two adjacent points which are within one-half mile of each other, etc. The scope of the transportation networks under consideration may be seen from

"Some British maps (group 1) calssify roads as:

1. Road, all weather motorable with milestones and bridge
2. Road, fair weather motorable

3. Road, unclassified
4. Pack-track
5. Foot-path."

In a transportation network, the ratio q/p of lines to points, i.e., of road segments to intersections, is called the "line density." On writing the equivalent of $p = e + n$, where e is the number of endpoints, called boundary points, and n is the number of non-endpoints, called internal points, various indices of density are tabulated for some transportation networks in Southeast Asia.

A similar approach is taken by Garrison [8] who studies transportation networks by taking the U. S. Interstate Highway System,

"In this paper the Interstate System is treated as a graph, and the usefulness of concepts from the theory of graphs is examined. Examination of the graph yields several measures which may be thought of as indices of relative location." Most transportation networks are planar graphs, but the presence of such complications as overpasses, tunnels, and viaducts may result in a nonplanar graph.

Organization Structure

Various concepts of graph theory, especially those involving connectedness and connectivity, can be given rather immediate interpretations in terms of the structure of interpersonal groups. For example, a cutpoint represents a liaison person, who constitutes the only contact between two or more otherwise disjoint social groups, and a bridge is a link between two groups which can be regarded as a meeting between two persons each of whom represents his group. Some cutpoints may be more essential for connectedness than others. A beginning toward the development of such a measure is made in a note with Ostrand [28]. A clique, being a maximal complete subgraph, is an important structural concept which has already proved useful in increasing the productivity of an organization by assigning people to work together in subgroups with a common task. These and similar ideas are developed in the report [13] of a lecture given at the Institute for Management Sciences.

We rushed into print some years ago in a paper with Ross [30] in which we proposed an algorithm exploiting the adjacency matrix of a graph for determining all its cliques. To set the record straight, that algorithm determines not only all the cliques of a graph, but sometimes a few other subgraphs as well. Correct algorithms for clique detection have subsequently been derived independently by several experts in computer programming.

Motivated originally by an anonymous article by C. N. Parkinson entitled "How to get seven employees to do the work of one," we defined in [12] the status of a person in an organization in terms of not only the number of people under him, but also how far under they were located. We then proposed a directionally dual concept, called contrastatus, which is the status a person would have in the inverted organization whose delegated authority relationships are obtained by turning the original organization chart upside down.

An article [17] in which a structural study is made of Mozart's celebrated opera "Cosi fan Tutte" has received rather wide notice, and has even been reprinted without permission of anyone in the Journal of Irreproducible Results. When the secretary of this journal retyped the original article, several misprints were introduced and subsequently an erratum was printed, in which the year of publication of the book [24] was changed from 1964 to 1965. The editor of this infamous journal then apologized for not having the correct year in the 1963 article. To the great surprise of the author, the Department (now Faculty) of Mathematics at the University of Waterloo thought that this article was written in homage to the distinguished graph theorist, W. T. Tutte, and posted it accordingly on their bulletin board. Subsequently, other articles have appeared in which graph theory is used to present the content analysis of the structure of a drama.

In another lecture being delivered at this symposium, Read [35], who founded the Computing Centre at the University of the West Indies, reports on computer algorithms for these and other graph theoretic concepts which were systematically and ingeniously derived with the assistance of his junior colleagues, Cadogan, King, Milner, and Parris.

Role Theory

The most recent and definitive work on role theory is contained in the book [2], edited by Biddle and Thomas, in which experts on various aspects of this theory expound on their specialties. The series of papers [25, 26, 27], written with Oeser, and [34] by Oeser and O'Brien, set forth a graph theoretic model for role theory. O. A. Oeser hopes and plans to write a book in which he proposes to show in detail that essentially all of the concepts of role theory can be made more elegant and less nebulous by appropriate effective exploitation of this model. This would be in direct contrast to the motto of sociology as stated by a scholar in Canberra, "It is vogue to be vague."

Consider three disjoint non-empty sets S_0, S_1, and S_2 as sets of points. Consider further a collection of directed graphs with the same point set S_0, and acyclic digraphs with points S_1 and S_2, respectively. We add to these three structures two bipartite graphs, one joining points of S_0 with those of S_1 and another whose sets of points are S_1 and S_2.

The interpretation of this multiply layered model in terms of role theory is as follows. The set S_0 stands for H, the collection of humans who are the people in the organization under consideration. The set S_1 is taken as P, the collection of positions or offices to be filled in the organization structure. The third set S_2 is T, the tasks to be accomplished by this organization. The collection of graphs and digraphs with point set H are the informal interpersonal relations among these people, including the relations of affect (like and dislike), informal power, respect, communication, etc. The superposition of all these relations, each of which can be regarded as a digraph provided there are no loops (such as self-communication, self-love), will be denoted for convenience by D_0. The organization chart is an acyclic digraph with point set P in which the lines from each point indicate those positions which are its immediate subordinates. The acyclic digraph on T is sometimes known as the precedence relation and shows which tasks must be done before others can be done.

The two bipartite digraphs are then D_{01} which exhibits the assignment of people to positions, and D_{12} which coordinates positions with the various tasks to be accomplished. This model is still only in its initial development, and calls for the addition of numbers on the directed lines, which changes the setting from digraphs to networks. These numbers can give such information as the amount of power, the percentage of time a person spends in each of his various positions, the cost of doing the different tasks, and so forth. The resulting information is then best handled, especially for large organizations, by means of matrices.

Markov Chains

There are very many applications of Markov chains in the social sciences. The two we mention are the treatment of brand loyalty in the paper [23] with Lipstein and the pioneering book [3] on learning theory by Bush and Mosteller. Although these are not primarily graph theoretic, the subject is related in the following way.

As noted in [23], a (finite, discrete, stationary) Markov chain M can be represented, after the initial probability vector has been invoked, by a directed graph (with directed loops permitted) together with a positive real number p_{ij} on each directed line from point v_i to v_j. When there is no such line, $p_{ij} = 0$. The one condition which the numbers p_{ij} must satisfy is that the sum of the numbers p_{ij} over all j is 1. The points v_i stand for the states or events of the chain. Each number p_{ij} gives the conditional probability that if the present event (the most recent event which has occurred) is v_i, then the next event will be v_j. Using the underlying digraph of a given chain, it is possible to classify its states as absorbing, ergodic, transient, etc.

In order to make an empirical study of brand switching and brand loyalty, the principal brands of a product (such as automobiles, toothpaste, coffee) may be taken as the events, together with one additional event for the miscellaneous brands. A panel of paid but impartial shoppers make reliable reports on their purchases during each buying period of time. After two such reports are gathered, the percent of those consumers who bought brand v_i the first period and brand v_j the second period can be calculated. Of course the diagonal entries p_{ii} of the transition probability matrix P_1 (using the first two reports) measure brand loyalty while the off-diagonal entries p_{ij} refer to brand switching. If the third report, when combined with the second report, results in a probability matrix P_2 which is "rather close" to P_1, then the process is fairly stationary. Such information as contained in the sequence of matrices P_1, P_2, P_3, ... has proved useful in the planning of advertising promotions and other allocations of the advertising budget. A modification of this formulation has also been valuable in making correct and rapid decisions regarding new product introductions, in particular, whether its promotion should be continued or dropped.

Bush and Mosteller express the events in their chain model for learning theory in the passage [3, p. 13]

"For even the simplest type of learning, such as bar-pressing by rats, we need two classes of responses, those which terminate in a relay closure indicating a bar press and those which do not. Although most of our analysis will deal with just two response categories, we shall define in general r mutually exclusive and exhaustive classes of responses."

These responses are the r events in the chain and the probability values refer to the conditional probability that if the present response is one of the alternatives a_i, then the next response will be a specified a_j. This chain need not be stationary, i.e., the conditional probabilities may vary with time.

Structural Balance

My own first exposure to graph theory, at the Research Center for Group Dynamics of the University of Michigan, arose when I was asked in 1950 to develop a mathematical model for group structure, with the word "group" standing for people (rather than an undefined term with a binary operation satisfying certain rules).

In a <u>signed graph,</u> each line is positive (solid line segments) or negative (dashed), as illustrated in Figure 1 with the four different signed triangles.

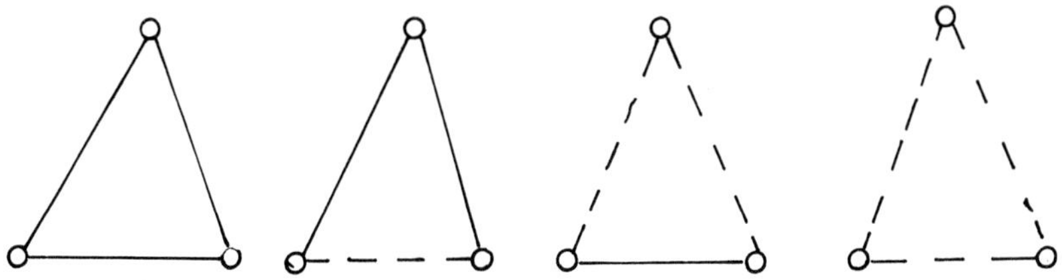

Figure 1. The signed triangles

Mathematically, signed graphs arise from the extension of the adjacency matrix of a graph from 0, 1 entries to include -1 as well. Psychologically, a signed graph depicts the simultaneous presence in a social group of a positive relation and its opposite negative relation, such as like and dislike.

The <u>sign of a cycle</u> is defined as the product of the signs of its lines. A signed graph is called <u>balanced</u> if all its cycles are positive. The first theorem characterizing structural balance appeared in [9] as mathematics and in [4] as part of a psychological theory.

<u>Theorem 1</u> A signed graph S is balanced if and only if there exists a partition of its point set V into one or two subsets such that every positive line joins two points of the same subset and every negative line joins two points of different subsets.

It turns out that this is merely a disguised form of König's Theorem (see [18, p. 18] on <u>bipartite graphs,</u> those in which the point set can be partitioned into two subsets with no line joining two points in the same subset.

<u>Theorem 2 (König)</u> A graph G is bipartite if and only if every cycle has even length.

The connection between Theorems 1 and 2 is rather immediate, and is obtained by taking the signed graph S of

Theorem 1, erasing all the positive lines, and calling the resulting structure (in which all lines are negative) a (bipartite) graph G.

Special cases of structural balance in signed graphs, such as local balance in which all cycles containing a given point are positive, are studied in [10]. Three different kinds of structural duality were defined in [11], namely (1) existential dual, the complement of a graph, (2) directional dual, the converse of a digraph, and (3) antithetical dual, the signed graph in which every line has altered sign. As noted in [11], S. Freud independently (and previously) discovered these and called them by such terms as economic and biological polarities.

A tendency of unbalanced social groups to achieve balance was hypothesized in [14]. The <u>line-deletion index</u> of balance in a signed graph S is the minimum number of lines whose removal from S results in balance. Similarly, the <u>line-negation index</u> is the minimum number of lines whose sign-change yields balance.

<u>Theorem 3</u>. In any signed graph, the line-deletion index and the line-negation index are equal.

No theorems are known concerning a corresponding point-deletion index. Incidentally, a formula for the number of balanced signed graphs was derived in [29], and a matrix criterion for balance in [15].

The concept of structural balance has become quite well known in psychological circles (also used for international relations in [16]), and much data has been collected to test this theory empirically. One of the findings so obtained is that the last triangle in Figure 1, the all negative one, does not necessarily represent an unstable social group. This development suggested the following generalization of balance obtained by dropping the words "one or two".

A signed graph was defined in [5] as <u>clusterable</u> if there is a partition of V into subsets such that every positive line joins two points of the same subset and no negative line does so.

Theorem 4. A signed graph is clusterable if and only if no cycle contains exactly one negative line.

One can make a transition from a clusterable signed graph S to its all-negative ordinary graph G by removing all the positive lines of S. This process changes the minimum number of clusters needed to partition the points of S to the chromatic number $\chi(G)$ of a graph G. By definition, $\chi(G)$ is the minimum number of colors needed for V so that no two adjacent points have the same color.

A complete n-coloring of V is one in which every pair of colors are assigned to some pair of adjacent points. An elementary homomorphism of G is the identification of two non-adjacent points; a homomorphism of G is a sequence of elementary homomorphisms. By a complete homomorphism of G is meant one from G onto some complete graph K_n. Of course there is a 1-1 correspondence between such complete homomorphisms and the complete n-colorings of G. Thus the minimum order of a complete homomorphism of G is its chromatic number $\chi(G)$. Analogously the maximum such order is called in [20] the achromatic number $\psi(G)$. The following interpolation theorem was obtained with Hedetniemi and Prins [21].

Theorem 5. Given any graph G, there is a complete n-coloring for every n such that $\chi(G) < n < \psi(G)$.

Let G be a labeled graph. Any $\chi(G)$-coloring of G induces a partition of the point set of G into $\chi(G)$ color classes. If $\chi(G) = n$ and every n-coloring of G induces the same partition of V, then G is called in [5, 22] uniquely n-colorable or simply uniquely colorable. For example, every even cycle is uniquely colorable and every odd cycle except the triangle is not.

A necessary condition for unique colorability was obtained with Cartwright [5].

Theorem 6. In the n-coloring of a uniquely n-colorable graph, the subgraph induced by the union of any two color classes is connected

This is an easy corollary of König's result, Theorem 2 above, which implies that a graph with $\chi = 2$ is uniquely colorable if and only if it is connected.

A classical result on coloring was extended to unique coloring with Hedetniemi and Robinson in [22].

Theorem 7. For all $n \geq 3$, there is a uniquely n-colorable graph which contains no subgraph isomorphic to K_n

Although the existence of a 5-chromatic planar graph is still open, a result of Hedetniemi (see [18, p. 140]) settles the problem for unique 5-colorability.

Theorem 8. No planar graph is uniquely 5-colorable.

Network Flows.

The original definitive work on this subject, which has proved most useful in applications to practical economic problems, is the book [7] by Ford and Fulkerson. This treatment has as its cornerstone the "max-flow min-cut theorem" (in which we do not here define the terms), which can be stated the the following form:

Theorem 9. In any network with positive values on its directed lines, the maximum flow from point u to point v is equal to the minimum cut capacity which separates u from v.

In his lecture on electrical networks, Minty [31] points out that his theorem (which is called the lemma of the colored arcs in the book [1] by Berge and Ghouila-Houri on network programming) has already been implicitly discovered by Ford and Fulkerson [7] as a very useful method of proof. It can be stated as follows:

Theorem 10. If G is a 2-connected graph in which one of the lines x is colored green and the remaining lines are arbitrarily colored red and blue, then there exists either a cycle containing x and no blue lines or a cocycle containing x and no red lines.

In a sense, all such results are at least intimately related and sometimes equivalent to the classical theorem of Menger (see [18, Chapter 5]):

Theorem 11. If u and v are nonadjacent points of a connected graph G, then the minimum number of points whose removal separates u and v equals the maximum number of point-disjoint paths joining u and v.

REFERENCES

1. C. Berge and A. Ghouila-Houri, Programming, Games & Transportation Networks, Methuen, London, 1965.
2. B. J. Biddle and E. J. Thomas (Eds.), Role Theory, Wiley, New York, 1966.
3. R. R. Bush and F. Mosteller, Stochastic Models for Learning, Wiley, New York, 1955.
4. D. Cartwright and F. Harary, Structural balance: a generalization of Heider's theory. Psychol. Rev. 63 (1956) 277-293.
5. D. Cartwright and F. Harary, On colorings of signed graphs, Elem. Math. 23 (1968) 85-89.
6. M. F. Dacey, A review on measures of contiguity for two and k-color maps. Spatial Analysis: a reader in statistical geography (B. Berry and D. Marble, eds.). Prentice-Hall, Englewood Cliffs, 1968, 479-490.
7. L. R. Ford and D. R. Fulkerson, Flows in Networks. Princeton University Press, Princeton, 1962.
8. W. L. Garrison, Connectivity of the interstate highway system. Papers and Proc. of the Regional Sci. Assoc. 6 (1960) 121-137.
9. F. Harary, On the notion of balance of a signed graph. Mich. Math. J. 2 (1953-54) 143-146.
10. F. Harary, On local balance and N-balance in signed graphs. Mich. Math. J. 4 (1955) 37-41.
11. F. Harary, Structural duality. Behav. Sci. 2 (1957) 255-265.

12. F. Harary, Status and contrastatus. Sociometry 22 (1959) 23-43.

13. F. Harary, Graph theoretic methods in the management sciences. Management Sci. 5 (1959) 387-403.

14. F. Harary, On the measurement of structural balance. Behav. Sci. 4 (1959) 316-323.

15. F. Harary, A matrix criterion for structural balance. Naval Res. Logist. Quart. 7 (1960) 195-199.

16. F. Harary, A structural analysis of the situation in the Middle East in 1956. J. Conflict Resolution 5(1961) 167-178.

17. F. Harary, 'Cosi fan Tutte'--a structural study. Psych. Reports 13 (1963) 466.

18. F. Harary, Graph Theory. Addison-Wesley, Reading, Mass., 1969.

19. F. Harary (Editor), Proof Techniques in Graph Theory. Academic Press, New York, 1969.

20. F. Harary and S. T. Hedetniemi, The achromatic number of a graph. J. Combinatorial Theory (1969), to appear.

21. F. Harary, S. T. Hedetniemi, and G. Prins, An interpolation theorem for graphical homomorphisms. Port. Math. 26 (1967), to appear.

22. F. Harary, S. T. Hedetniemi, and R. W. Robinson, Uniquely colorable graphs. J. Combinatorial Theory 6 (1969) 264-270.

23. F. Harary and B. Lipstein, The dynamics of brand loyalty: a Markovian approach. Operations Res. 10 (1962) 19-40.

24. F. Harary, R. Z. Norman, and D. Cartwright, Structural Models: an introduction to the theory of directed graphs. Wiley, New York, 1965.

25. F. Harary and O. A. Oeser, A mathematical model for structural role theory, I. Human Relations 15 (1962) 89-109.

26. F. Harary and O. A. Oeser, A mathematical model for structural role theory, II. Human Relations 17 (1964) 3-17.

27. F. Harary and O. A. Oeser, Role structures: a description in terms of graph theory, Role Theory, Wiley, New York, 1966.

28. F. Harary and P. A. Ostrand, How cutting is a cutpoint? Combinatorial Structures and Their Applications (R. K. Guy, ed.) Gordon and Breach, New York. (to appear).

29. F. Harary and E. M. Palmer, On the number of balanced signed graphs. Bull. Math. Biophysics 29 (1967) 759-765.

30. F. Harary and I. C. Ross, A procedure for clique detection using the group matrix. Sociometry 20 (1957) 205-215.

31. G. Minty, Graph theory and electrical networks, Graph Theory and its Applications (B. Harris, ed.) Academic Press, New York 1970.

32. J. D. Nystuen, Analysis of Geographic and Climatic Factors in Coastal Southeast Asia, Section VII, Roads and Communications, University of Michigan Report No. 04231-1-F, 1962.

33. J. D. Nystuen and M. F. Dacey, A graph theory interpretation of nodal regions, Papers and Proc. of the Regional Sci. Assoc. 7 (1961) 29-42.

34. O. A. Oeser and G. O'Brien, A mathematical model for structural role theory, III, Human Relations 20 (1967) 83-97.

35. R. C. Read, Graph theory algorithms, Graph Theory and its Applications (B. Harris, ed.) Academic Press, New York, 1970.

36. R. L. Wilder, Trends and social implications of research, Bull. Amer. Math. Soc. 75 (1969) 891-906.

Research supported by Grant No. MH 10834-06 from the National Institute of Mental Health.

The Mystery of the Heawood Conjecture

J. W. T. YOUNGS

"A remarkable fact connected with the four color problem is that for surfaces more complicated than the plane or the sphere the corresponding theorems have actually been proved, so that, paradoxically enough, the analysis of more complicated geometrical surfaces appears in this respect to be easier than that of the simplest cases."

The above sentence starts a paragraph in a book by Courant and Robbins ([1] p. 248), which first appeared in 1941, and the sentence has remained unchanged through later editions. In fact, the statement was false when written and remained so for almost 30 years. "For surfaces more complicated than the plane or the sphere the corresponding theorems" had, at that time, not been proved. The corresponding theorems involve determining the chromatic number of surfaces of positive genus. If S_p is such a surface, with genus p, then the Heawood conjecture [3] is that $\chi(S_p)$, the chromatic number of S_p, is $H(p)$, the integral part of $(7 + \sqrt{1+48p})/2$, and the conjecture was confirmed only in 1968 (see [7]).

The object of these remarks is not, however, to reproach Courant and Robbins for their oversight, because they merely stated a belief held by segments of the mathematical community for many years (see [7]). An initial mystery, then, is how did it happen that such a myth invaded mathematical folklore, finally to be accepted as truth? The answer can probably only be a guess, but it does not appear to be an entirely unreasonable explanation that it was the result of reckless extrapolation from the well known and highly elegant proof that the chromatic number of S_1, the torus, is 7. The reader is invited to submit alternative hypotheses.

A true mathematical mystery occurs in the history of the problem. The Heawood conjecture was stated in 1890 and Heffter [4] made significant contributions almost immediately in 1891. Why then did it take the best part of a century before the matter was settled? After all, both Heawood and Heffter lived long, active mathematical lives (each into his tenth decade) and it is unthinkable that they did not consider the matter from time to time. Moreover, there were others involved in map coloring problems during the intervening years. Here it seems we are on more solid ground in trying to provide an answer. Somehow investigators missed two rather simple techniques which are of fundamental importance in the present solution to the problem. The techniques have nothing in common, but their union suffices to settle the matter. More will be said about this later.

A useful geometric model for S_p is a sphere with p attached handles. A <u>map</u> $\mathfrak{M}$ on S_p is a cellular decomposition of S_p. Each 2-cell of $\mathfrak{M}$ is called a <u>country</u> and 2 countries are adjacent if some 1-cell of $\mathfrak{M}$ is a face of both of them. A <u>coloring</u> of $\mathfrak{M}$ is an assignment of a color to each country so that adjacent countries are not assigned the same color. The number of colors used in a coloring of $\mathfrak{M}$ is called the chromatic number of the coloring, and $\chi(\mathfrak{M})$, the <u>chromatic number of</u> $\mathfrak{M}$ is the minimum chromatic number for all colorings of $\mathfrak{M}$. The <u>chromatic number of</u> S_p, designated by $\chi(S_p)$, is the maximum of $\chi(\mathfrak{M})$ for all maps $\mathfrak{M}$ on S_p. At first sight there is no reason to believe that $\chi(S_p)$ exists (i. e., is finite), however Heawood showed that

$$\chi(S_p) \le H(p) \quad \text{if } p > 0 . \tag{1}$$

The <u>Heawood conjecture</u> is that

$$\chi(S_p) = H(p) \quad \text{if } p > 0 . \tag{2}$$

The proof of the conjecture is in 12 cases depending upon the residue class of $H(p)$ modulo 12, and case k concerns (2) for all $p > 0$ such that $H(p) \equiv k \pmod{12}$, $k = 0, \ldots, 11$.

A rough outline of the argument is presented below, but it is convenient to start with some definitions and a notation.

A cubic graph is one with the property that there are three arcs incident at each vertex, and a regular map is a map whose 1-skeleton is a cubic graph. The notation $\mathfrak{m}_n$ is reserved for a regular map with exactly n countries, each adjacent to all the others.

For each $k = 0, \ldots, 11$ the argument which confirms the the Heawood conjecture is as follows. Select any $s \geq 0$, and let $n = 12s + k$. Define p to be the smallest integer q such that $H(q) = n$, and construct a map $\mathfrak{m}_n$ on S_p. Thus $\chi(S_p) \geq \chi(\mathfrak{m}_n) = n = H(p)$ and (1) implies (2) if $p > 0$. With k fixed, such a result for each $s \geq 0$ is sufficient to establish (2) in case k (see [9]).

Because of the Euler formula, in certain cases ($k = 0, 3, 4, 7$) the map $\mathfrak{m}_n$ on S_p will have the additional property that each pair of countries has exactly one 1-cell as a common face. These are called the regular cases. If $k = 1, 2, 5, 6, 8, 9, 10, 11$ then the Euler formula rules out the possibility of constructing such regular maps on S_p. Here one first constructs a regular map with $n = 12s + k$ countries on a surface S_r of suitably lower genus. The map is regular but certain countries are not adjacent. This construction is called the regular part of the problem and requires the theory of vortices [10]. The remaining problem, roughly speaking, is to add $(p-r)$ suitably placed handles to S_r and enlarge the non-adjacent countries so as to obtain the missing adjacencies, thus arriving at a map $\mathfrak{m}_n$ on S_p. This is called the irregular or additional adjacency part of the problem, and Ringel has devised techniques for dealing with the matter (see [6] and [8]).

The two techniques used for resolving the Heawood conjecture are therefore the theory of vortices and the theory of additional adjacencies. Neither of these was available to Heffter. Nevertheless, Heffter made an interesting attack on case 7 and this note is to be considered a tribute to him as well as a critical analysis of his results.

In view of the fact that case 7 is one of the regular

cases, a special form of the theory of vortices is employed and the technique for obtaining additional adjacencies is not needed. The special form is the theory of current graphs with rotation, and was discovered by Gustin [2]. This was a spectacular advanced made in the spring of 1962. (For other historical comments, see [7] and [10].) We propose to cast Heffter's results in the language of current graphs and note the results. This exercise will reveal why he failed to solve case 7 completely.

I. Heffter's results.

We are concerned only with case 7; that is with the confirmation of (2) for those p such that $H(p) \equiv 7$ modulo 12. A direct computation shows that the smallest integer q such that $H(q) = 12s + 7$ is $12s^2 + 7s + 1$. In accord with the outline of the argument presented above, we propose to confirm the Heawood conjecture in case 7 by constructing a map $\mathfrak{M}_n$ on S_p with $n = 12s + 7$ and $p = 12s^2 + 7s + 1$ for each $s \geq 0$.

As an introduction it pays to study Heffter's solution for the torus as depicted in figure 1.

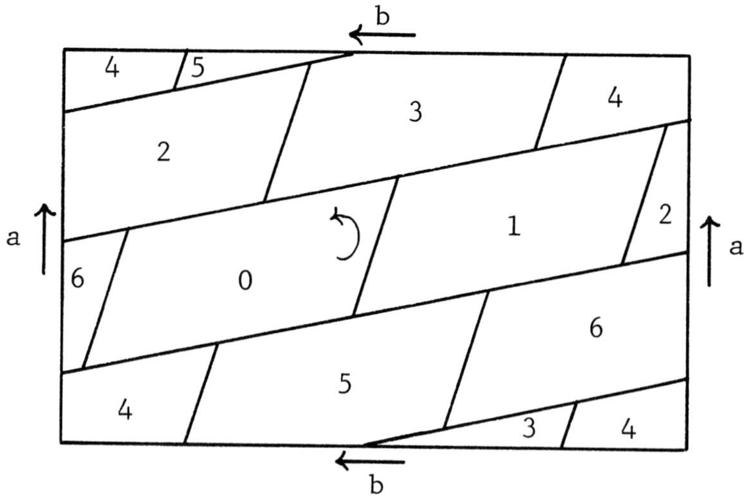

Figure 1

Record all the countries adjacent to country 0 in the cyclic order given by the displayed orientation on the torus to get

$$0.\quad 1\ 3\ 2\ 6\ 4\ 5\ .$$

Doing the same for countries $1, \ldots, 6$ (and repeating the permutation around 0 for completeness) we get an array

$$\begin{array}{l} 0.\ \ 1\ 3\ 2\ 6\ 4\ 5 \\ 1.\ \ 2\ 4\ 3\ 0\ 5\ 6 \\ 2.\ \ 3\ 5\ 4\ 1\ 6\ 0 \\ 3.\ \ 4\ 6\ 5\ 2\ 0\ 1 \\ 4.\ \ 5\ 0\ 6\ 3\ 1\ 2 \\ 5.\ \ 6\ 1\ 0\ 4\ 2\ 3 \\ 6.\ \ 0\ 2\ 1\ 5\ 3\ 4 \end{array} \tag{3}$$

called a *scheme* Σ_7. It describes the "adjacency pattern" of $\mathfrak{M}_7$ on S_1.

An important property of (3) is that a rule called R^* by Ringel [5] holds. R^* is a rule concerned with any two different rows i and k in the scheme and states that

$$i.\ \cdots \alpha\ k\ \beta \cdots \quad \underline{\text{implies}} \quad k.\ \cdots \beta\ i\ \alpha \cdots$$

(The fact that (3) satisfies R^* is most easily checked by glancing at figure 1 from which it came.)

From our point of view, the important fact is that, as Ringel [5] has shown (and Heffter used without explicit proof) the converse of the situation holds. To be more precise, for any n, a *scheme* Σ_n is an array of n rows, $0, \cdots, (n-1)$, where each row i is a cyclic permutation of the elements $0, \cdots, \hat{i}, \cdots, (n-1)$. Ringel has shown that:

If a scheme Σ_n *satisfies* R^* *then there is a surface* S_p *with a regular map* $\mathfrak{M}_n$ *on it; in fact, each pair of different countries in* $\mathfrak{M}_n$ *share exactly one 1-cell.*

Let us examine the cellular decomposition of S_p given by $\mathfrak{M}_n$. If α_i is the number of i-cells, $i = 0, 1, 2$, then $\alpha_2 = n$ because there are n countries. The adjacencies between countries of $\mathfrak{M}_n$ show that $\alpha_1 = n(n-1)/2$. Finally, the regularity of the map implies that $3\alpha_0 = 2\alpha_1$. The Euler characteristic E of S_p is $\alpha_0 - \alpha_1 + \alpha_2$ and $p = 1 - E/2$. Hence $p = (n-3)(n-4)/12$.

In our application $n = 12s + 7$ and therefore $p = 12s^2 + 7s + 1$. Hence, referring to the first paragraph of this section, the Heawood conjecture is confirmed in case 7 if we can find a scheme Σ_n satisfying R^* for each $n = 12s + 7$, $s = 0, 1, 2, \cdots$.

A scheme Σ_n satisfying R^* is said to be an orientable scheme for $\mathfrak{M}_n$.

A further glance at (3) is rewarding. If we interpret $0, \cdots, 6$ as the elements of Z_7, the additive group of integers modulo 7, then we observe that all the information is carried in row 0. The other rows are manufactured by the

Additive Principle. To obtain row i, add i to each element of the permutation in row 0 without changing the order.

A scheme obtained from row 0 by the additive principle is said, in the general theory, to be of index 1. (See [2] and [10].)

Heffter showed that to get orientable schemes of index 1 for $\mathfrak{M}_n$ using Z_n it is necessary for $n \equiv 7 (\mathrm{mod}\ 12)$. He thus dealt exclusively with case 7 and was able to obtain such schemes for all integers s such that if $q = 4s + 3$, then the order of 2 in the multiplicative group of integers modulo q is

(4)
a) $q - 1$, or
b) $(q - 1)/2$.

Moreover, he showed that his method of constructing row 0 could not be extended to other integers n congruent to 7 modulo 12.

Unfortunately it is not known to this day if there is

an infinite set of integers s with property a) or b). Hence, because of an unsolved problem in number theory, we do not know whether or not Heffter succeeded in resolving the Heawood conjecture for an infinite collection of surfaces.

Before examining Heffter's results in detail it is useful to have a further look at rule R^*. Given a scheme Σ_n, it satisfies Δ^* if for $i \neq k$,

$$i.\ \cdots \alpha\ k \cdots \quad \underline{\text{implies}} \quad k.\ \cdots i\ \alpha \cdots .$$

This has been called the rule of triangles in [8] where it is shown, in a few lines, that it is equivalent to R^*.

<u>Lemma 1.</u> If a scheme Σ_n is of index 1 and $0.\ \cdots a\, b \cdots$ implies $b.\ \cdots 0\ a \cdots$ then the scheme satisfies Δ^* (and therefore R^*).

Proof. Suppose $i \neq k$ and

$$i.\ \cdots \alpha\ k \cdots .$$

Since the scheme has index 1, this implies that

$$0.\ \cdots \alpha\text{-}i\ k\text{-}i \cdots ,$$

and by hypothesis

$$k\text{-}i.\ \cdots 0\ \alpha\text{-}i \cdots$$

Since the scheme has index 1 we obtain

$$k.\ \cdots i\ \alpha \cdots .$$

An immediate consequence is

<u>Lemma 2.</u> If a scheme Σ_n is of index 1 and $0.\ \cdots a\ b \cdots$ implies $0.\ \cdots -b\ a\text{-}b \cdots$ then the scheme satisfies Δ^* (and therefore R^*).

The statement of Lemma 1 is easier to remember but Lemma 2 is more useful in that it is a condition on row 0 alone. (In this connection see also Ringel [6].)

We now return to Heffter's results. If $s = 0, 4, 14, \cdots$

then $q = 4s + 3$ is of type a) - see (4) - and Heffter provides a method for obtaining row 0 of an index 1 orientable scheme for $\mathfrak{M}_n$ with $n = 12s + 7$. We shall call these type a) solutions. Type b) solutions occur if $s = 1, 5, 11, \ldots$. Here q is of type b).

For $s = 1$ his row 0 for $\mathfrak{M}_{19}$ using Z_{19} is

(5) $$0.\ \ 1\ \ 3\ \ 7\ \ 6\ \ \bar{5}\ \ 4\ \ \bar{3}\ \ \bar{2}\ \ 9\ \ 5\ \ \bar{8}\ \ 2\ \ \bar{1}\ \ \bar{7}\ \ \bar{4}\ \ \bar{9}\ \ 8\ \ \bar{6},$$

where $\bar{k}$ is defined to be $-k$ and the elements of $Z_{19}\setminus 0$, the non-zero elements of Z_{19}, are represented in the form $\pm k$, $k = 1, \ldots, 9$.

For $s = 2$ his row 0 for $\mathfrak{M}_{31}$ using Z_{31} is

(6) $$\begin{array}{l} 0.\ \ 1\ \ 3\ \ 7\ \ 15\ \ 9\ \ \overline{12}\ \ \overline{14}\ \ 13\ \ 5\ \ 11\ \ 10\ \ \bar{9}\ \ 6\ \ \bar{5}\ \ 8\ \ \bar{7}\ \ \bar{4} \\ \quad 14\ \ 2\ \ \bar{1}\ \ \overline{11}\ \ \bar{6}\ \ \overline{15}\ \ \bar{8}\ \ \overline{13}\ \ 4\ \ \bar{3}\ \ \bar{2}\ \ 12\ \ \overline{10} \end{array}$$

It is a laborious exercise to check that these rows give orientable schemes. The reader is asked not to do so because in the next section the matter will become practically obvious. We do not intend to show how Heffter's methods enabled him to obtain the rows. Our object is to cast his results in the context of the general theory of current graphs and rotations.

II. Current graphs and rotations

A graph K is understood to have neither loops nor multiple arcs. If A is an arc in the graph K it has two orientations on it; if α is regarded as A with one of these orientations, then $-\alpha$ represents A with the other orientation. Each oriented arc has an initial and a terminal vertex determined by the orientation. Reversing the orientation interchanges the vertices. (Given an oriented arc α the same arc without orientation is designated by $|\alpha|$ or $|-\alpha|$.) Thus each arc in K gives rise to two oriented arcs. The totality of oriented arcs in K is designated by K^*.

A current graph is a triple (K, G, λ) where K is a graph, G is an abelian group and λ is a transformation

$$\lambda : K^* \to G\setminus 0$$

subject to the condition that

(7) $$\lambda(-\alpha) = -\lambda(\alpha) \ .$$

We give several examples of representations of current graphs in figures 2 to 5 . In each the points A and B (if applicable) are identified. Each arc in the examples is displayed with one of its two possible orientations by the use of an arrow, and the element g of G associated with the oriented arc by λ appears near the arrow. The arc with opposite orientation is not shown; it is understood that, in view of (7) , the associated element is -g.

It is clear that figures 4 and 5 generalize for all $G = Z_{12s+7}$.

Given a current graph (K, G, λ) and $\alpha \in K^*$ the element $\lambda(\alpha)$ of $G \setminus 0$ is called the current on α . A current graph is said to satisfy Kcl (= Kirchhoff's current law) at a vertex if the sum of the currents on the oriented arcs which terminate at the vertex is 0 in the group G . Note that in the current graphs of figures 2 to 5 Kcl is satisfied at all the vertices. The reader's attention is drawn to figure 2 and the extremal vertices at the lower level. At the first of these the sum of the "inward flowing" currents is -19 , at the second 19, and Kcl holds because $\pm 19 = 0$ in $G = Z_{19}$.

Given a graph K a rotation at a vertex of K is a cyclic permutation of the oriented arcs which terminate at that vertex. A rotation r on K is a rotation at each vertex of K. A rotation on K is therefore a 1-1 transformation r from K^* onto itself.

A rotation r on K induces a second 1-1 transformation ρ from K^* onto itself defined by

$$\rho(\alpha) = -r(\alpha) \ , \quad \alpha \in K^* \ .$$

For any $\alpha \in K^*$ consider the sequence of oriented arcs

(8) $$\alpha, \ \rho(\alpha), \ \rho^2(\alpha), \ \cdots, \ \rho^k(\alpha) \ .$$

This is what is usually called an oriented path from the initial vertex of α to the terminal vertex of $\rho^k(\alpha)$. Because of the character of ρ there is a smallest integer k for

$G = Z_{19}$

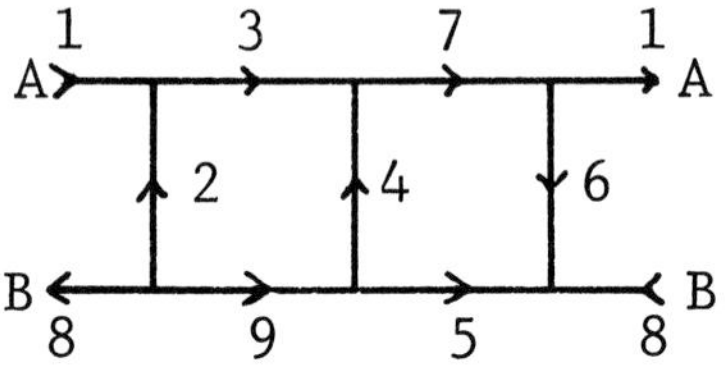

Figure 2

$G = Z_{31}$

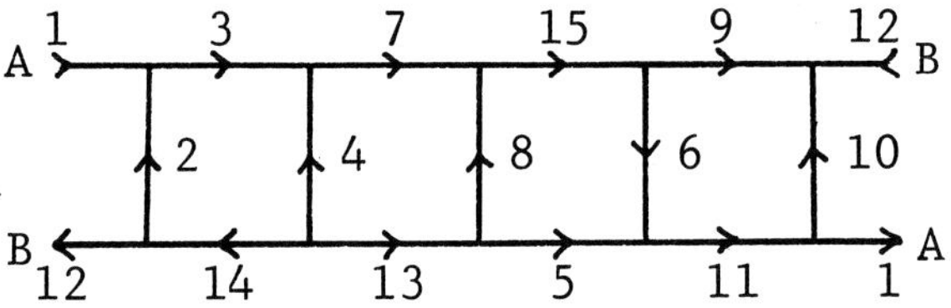

Figure 3

$G = Z_{31}$

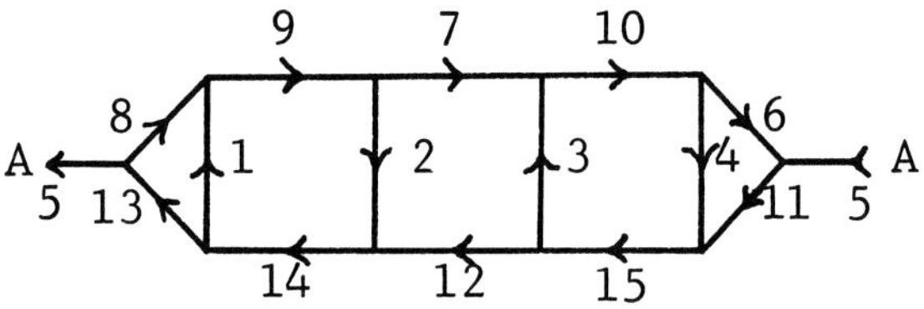

Figure 4

$G = Z_{43}$

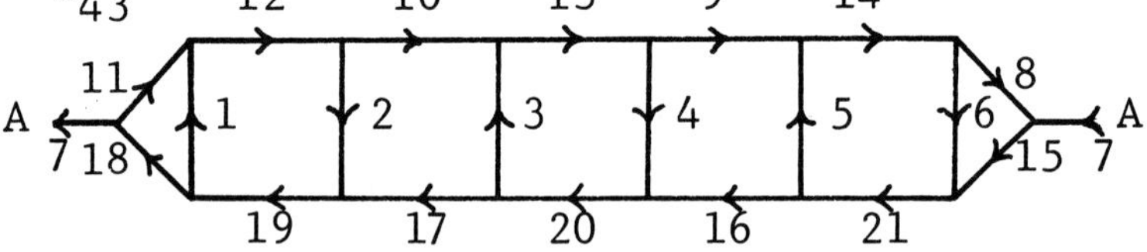

Figure 5

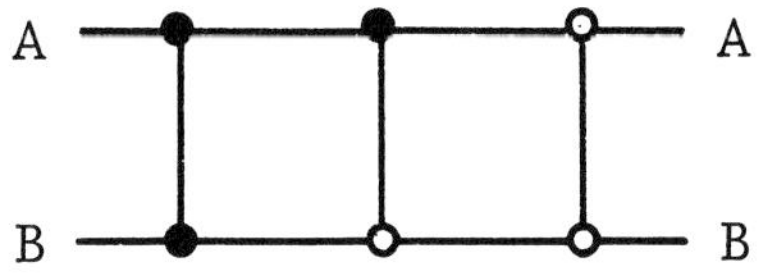

Figure 6

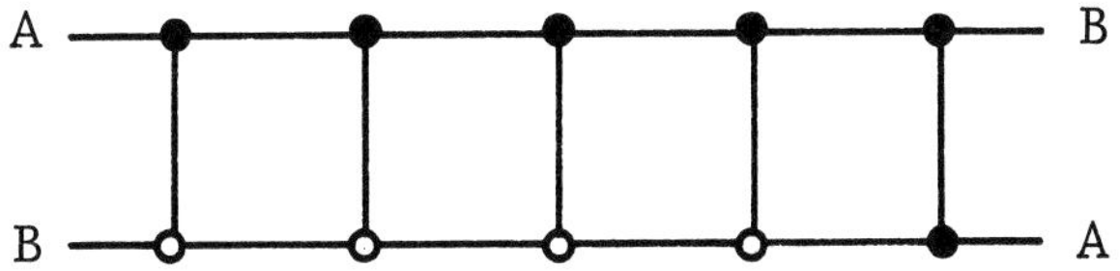

Figure 7

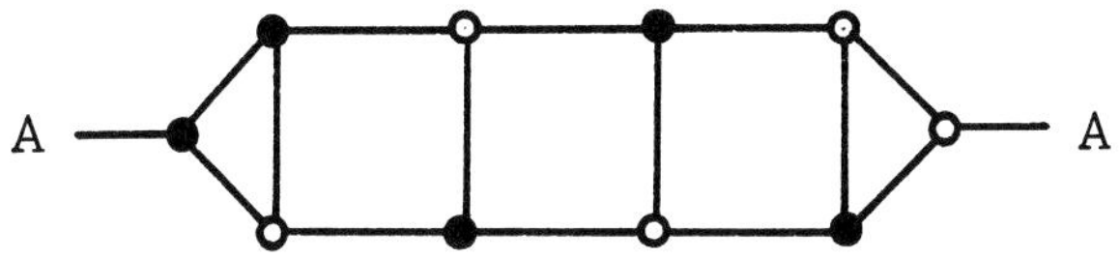

Figure 8

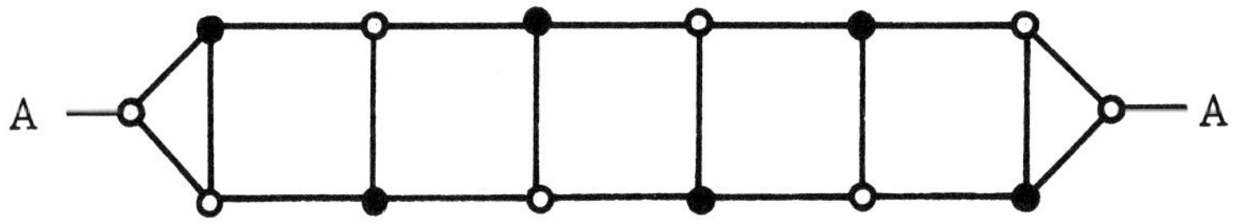

Figure 9

which $\alpha = \rho^{k+1}(\alpha)$. For this k the oriented path (8) is called a circuit induced by r and α, and (8) is then considered a cyclic permutation.

In this application we are particularly interested in rotations r which have the property that r induces the same circuit for all $\alpha \in K^*$. This means that each $\alpha \in K^*$ appears in the circuit exactly once, and r is said to induce a single circuit.

A standard notation in representing a graph K with rotation r is to draw it on the plane so that the rotation r at a vertex v is given by a clockwise, or counterclockwise, reading of the oriented arcs which terminate at v. If a clockwise (counterclockwise) reading is required the vertex is displayed as a small filled-in (empty) circle.

As examples of graphs with rotation we offer figures 6 to 9 in which the geometric structure is the same as figures 2 to 5 respectively.

The generalization of figure 6 to a ladder with an odd number of rungs is not entirely obvious. The rotation is clockwise at both ends of the first rung and counterclockwise at both ends of the last. In between the rotation is clockwise on top and counterclockwise below. The pattern of figure 7 is extensive enough so that the reader should have no difficulty in generalizing the situation to a ladder with any odd number of rungs. For each of these graphs the rotation induces a single circuit. As to figures 8 and 9, we deal with graphs having an even number of rungs. Figure 9 is not a generalization of figure 8, because the first vertex in the former has counterclockwise rotation. Hence, for each graph of the type shown in figures 8 and 9, with an even number of rungs, there are two sets of rotations (differing only at the first vertex) which follow the displayed patterns. In both cases the rotation on the graph induces a single circuit.

It is time to turn our attention to a basic theorem from the general theory [10].

Theorem. If (K, G, λ) is a current graph with the property that:

1. K is cubic,
2. $\lambda: K^* \to G\setminus 0$ is 1-1 and onto,
3. Kcl holds at each vertex,

and

4. There is a rotation r on K that induces a single circuit;

then the succession of currents on the oriented arcs of the circuit is row 0 of an index 1 orientable scheme for $\mathfrak{M}_n$ where n is the order of G.

Proof. The single circuit must contain every $\alpha \in K^*$. Since λ is onto, for every $g \in G\setminus 0$ there is an arc α such that $\lambda(\alpha) = g$. Since λ is 1-1 there is no arc $\beta \neq \alpha$ such that $\lambda(\beta) = g$. Therefore the succession of currents on the circuit is a permutation of $G\setminus 0$, and hence is a possible row 0 for the scheme. Suppose

$$0. \quad \cdots \ a \ b \cdots$$

and suppose α and β are oriented arcs in K such that $\lambda(\alpha) = a$ and $\lambda(\beta) = b$. Then the terminal vertex of α is the initial vertex of β and $\rho(\alpha) = \beta$. Since K is cubic we may represent the situation by figure 10 where x is unknown.

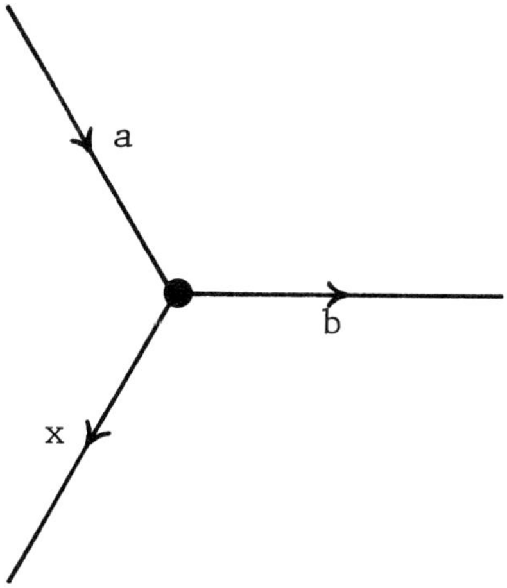

Figure 10

But Kcl is satisfied at the vertex, hence $a-b-x = 0$ or $x = a-b$.

But row 0 must also contain $\cdots -b\ x \cdots$; that is, $\cdots -b\ a-b \cdots$ and now by Lemma 2 the scheme is orientable. (This idea is so simple that it is difficult to see how Heffter missed it.)

III. Heffter's results in new form

For our present purposes it is a converse of the theorem which is interesting [10].

If there is an index 1 orientable scheme for $\mathfrak{m}_n$ then there is a current graph (K, G, λ) with properties 1, 2, and 3 of the theorem and a rotation r on K with property 4 such that the row 0 produced by the theorem is the row 0 of the given scheme.

Consequently we can capture Heffter's results. As illustrations of this capture, the reader should put together figures 2 and 6 to get Heffter's row 0 for $s = 1$ displayed in (5). Figures 3 and 7 give the row 0 of (6) for $s = 2$. In fact figures 4 and 8 give another solution for $s = 2$, while figures 5 and 9 a solution for $s = 3$. These last solutions are due to Ringel [6] without using this theory. We shall refer to the matter at the end of the paper.

It pays to examine the new form of Heffter's results more closely.

If $s = 1$, then $q = 4s+3 = 7$ and the successive powers of 2 in Z_7 are 2, 4, 8 = 1. If we represent these in the form $2k$ where $0 < |2k| < 7$ we obtain

$$2 \quad 4 \quad \bar{6} \quad . \tag{9}$$

Now look at the ladder (called cylindrical because the identification is without a "twist") of figure 2. The succession of currents on the rungs, if we regard the positive direction as upward, is precisely (9). If the currents at the left end are denoted in general terms, then the situation is as in figure 11.

All the other currents are determined in succession, moving from left to right, by Kcl and the rung currents (9). It is useful to observe that in each "box" the algebraic sum of the two horizontal currents is q to the left. The ladder current is defined to be this algebraic sum.

If $s = 2$, then $q = 11$ and the successive powers of 2 in Z_{11} when represented as in (9) are

(10) $$2\ 4\ 8\ \bar{6}\ 10\ \bar{2}\ \bar{4}\ \bar{8}\ 6\ \overline{10}\ .$$

The ladder of figure 3, called of Möbius type because of the "twist", has the first 5 elements of (10) as rung currents with directions up except for 6. Moreover, the currents at the left end are as in figure 11. As in the case $s = 1$, of course, figure 11 and the rung currents determine the complete current graph subject to Kcl.

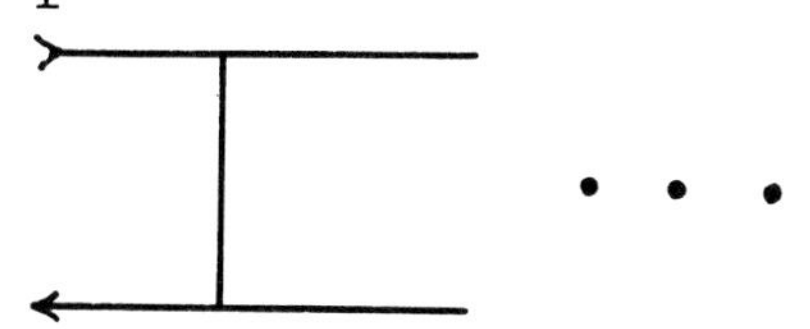

Figure 11

In a nutshell, what Heffter did, in terms of current graphs, was as follows:

1. If s is such that $q = 4s+3$ is of type a) or b) -see (4) - write down the first $2s+1$ successive powers of 2 in Z_q representing them in Z_q as $2k$ where $0 < |2k| < q$.

2. Construct a horizontal ladder with $2s+1$ rungs but do not decide in advance whether it is to be cylindrical or Möbius type. Place the $2s+1$ currents of 1 in order and in positive form on the rungs, orienting the rung down if and only if the current is negative. These currents are now regarded as elements of Z_{12s+7}.

3. With the currents at the left, as in figure 11, use Kcl to place currents on the horizontal arcs. For convenience represent these currents in the form $1, \ldots, 6s+3$.

Once steps 1, 2 and 3 have been taken it is a theorem of Heffter (in current graph terms) that the final diagram is a current graph with a cylindrical (Möbius type) ladder in case q is of type b (type a); moreover the current graph has the first three properties of the theorem. In regard to an appropriate rotation, statements have already been made while considering figures 6 and 7 .

This is Heffter's solution, and we see that in virtue of the theorem our hands are severely tied by the preordained manner of determining the rung currents. With this restriction Heffter's result is final and complete.

We propose to try to find a Heffter type solution where we limit the $2s+1$ rung currents to the numbers $2, 4, 6, \ldots, 4s+2 = q-1$, as he did, but do not predetermine their distribution or direction.

To do this we first propose to examine the following

Generalized Problem. Consider the ladder K_t of figure 12 with t rungs where the ladder may be of type $\underset{\sim}{C}$ (cylindrical) or $\underset{\sim}{M}$ (Möbius).

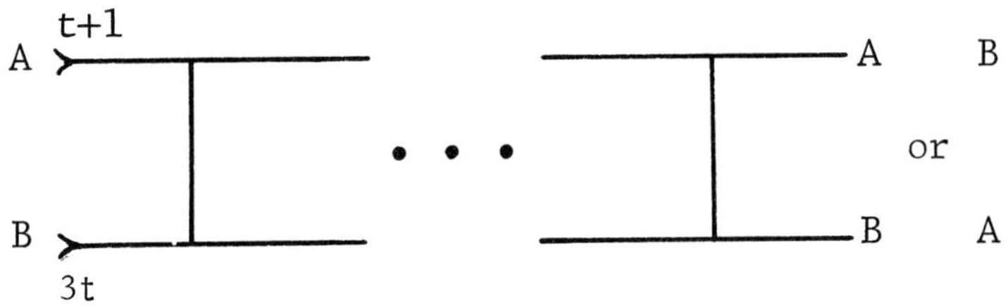

Figure 12

The currents already on K_t are from Z_{6t+1}.

Fill in the rest of the ladder so as to get a current graph $(K_t, Z_{6t+1}, \lambda_t)$ with the following properties:

1) $\lambda_t: K_t^* \to Z_{6t+1}$ is 1-1 and onto ,

(11) 2) Kcl holds at each vertex,

3) $1, \cdots, t$ appear on the rungs.

(Note that in the generalized problem we disregard rotation.)

If this can be done, then on multiplying each current by -2 (which is relatively prime to $12s + 7$) and reflecting the graph about its horizontal axis we shall obtain a Heffter type solution whenever t is odd. For example $t + 1$ will become $-2t-2$ or $2t+2$ pointing left. After reflection it will appear at the bottom, and if $t = 2s+1$ will be $4s+4 = q+1$ pointing left. Similarly $3t$ will become $-6t = 1$ pointing right at the top. This will agree with figure 11 and the generalized problem is much more convenient than trying to find Heffter type solutions directly.

There will be the further aesthetic advantage that Kcl at each vertex will not only hold in Z_{6t+1}, it will hold in Z. (See the remarks about figure 2.)

IV Solution of the generalized problem

Condition 1) of the problem means that the currents

$$(12) \qquad t+1, \ldots, 3t$$

must appear on the $2t$ horizontal arcs. (Note that the "half" arcs at the left have their "other halves" at the right in some vertical order.)

The currents $t+1$ and $3t$ appear pointing to the right at the beginning of the diagram. Hence the ladder current is $4t+1$ to the right. This means that if $t+k$ is a current on the horizontal arc, it must point to the right, and $3t-k+1$ must be its companion current on the other horizontal arc in the same box, and must also point to the right. This uniformity of current direction on the sides of the ladder is most helpful - note, in contrast, figures 2 and 3. We must therefore consider the "horizontal" currents (12) in pairs

$$P_k: t+k, \quad 3t-k+1$$

for $k = 1, \ldots, t$. For fixed k the two currents in P_k will appear in the same box, and both will point to the right.

A major problem in this case is that it is impossible, within reasonable limitations of space, to draft enough current graphs for successive t's to make the continuation for larger t's obvious. A terse codification of the results is essential and we recall the technique of zigzag diagrams which has proved very useful in the past. (See [8] and [11].)

We propose to use it here. As a matter of fact, the technique can also be used as a nomogram to obtain solutions, and the solutions to be presented here were so discovered. In the interests of brevity, however, we shall use the technique simply to record the results.

To this end fix t and consider t equally spaced points on a horizontal k-axis, naming them $1, \ldots, t$. The point at k stands for the pair P_k. On the same axis define a <u>mass function</u>,

$$m = 2(t-k)+1$$

for $k = 1, 3/2, 2, 5/2, \ldots, t-3/2, t-1, t-1/2, t$. So far the whole situation can be conveniently displayed in figure 13 for $t = 7$ and 8. The numbers taken on by the linear function m are displayed on the k-axis for $k = t - 1/2, t - 3/2, t - 5/2 \ldots$ at these values of k. The display is ended at that k for which $m = t-1$ if t is odd, and $m = t$ if t is even.

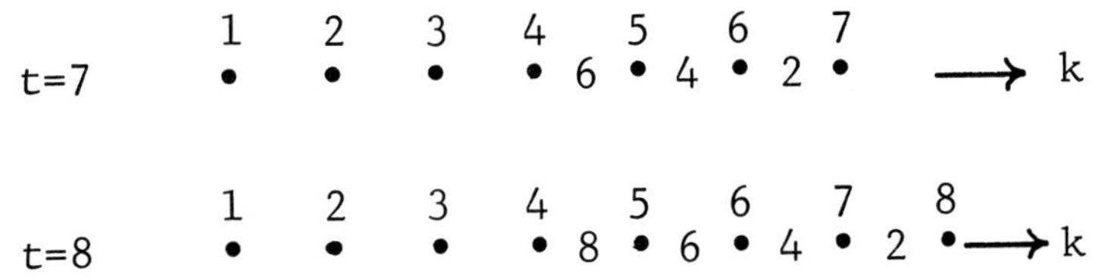

Figure 13

We shall show first that the zigzag diagram in figure 14 is a solution to the generalized problem for $t = 7$. Note that the zigzag has the following properties.

Z1. The zigzag starts and terminates at $k = 1$ (representing P_1) with one "riser" under each $k = 2, \ldots, 7$.

Z2. The "labels", recorded to the right of the successive steps, are $1, \ldots, 7$, and are the same as the currents in item 3 of the generalized problem.

Z3. A label in ordinary type is the length of the adjacent step.

Z4. A label in boldface type ($\underset{\sim}{6}$ and $\underset{\sim}{7}$) is the value of the mass function at the midpoint of the adjacent step.

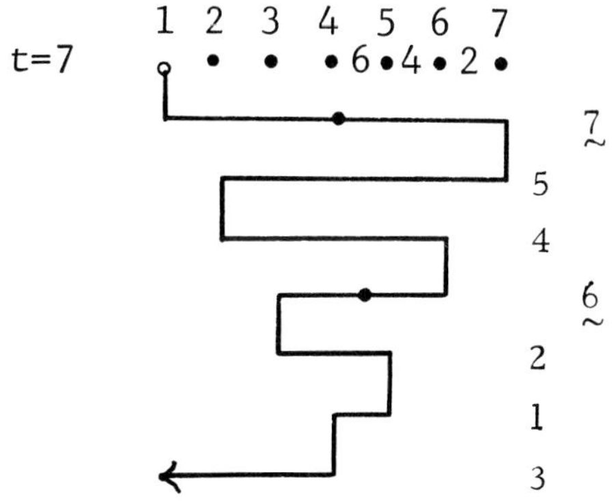

Figure 14

Now consider a verbal description of the zigzag. It starts at 1 (representing P_1) and goes to 7 (representing P_7) by a step labelled $\underset{\sim}{7}$; then to 2 by a step labelled 5; · · · finally from 4 to 1 by a step labelled 3. Using the obviously suggested notation we can abbreviate the discription of the zigzag to read as an alternating sequence of P's and labels:

$$P_1 \underset{\sim}{7} P_7 \; 5 \; P_2 \; 4 \; P_6 \underset{\sim}{6} P_3 \; 2 \; P_5 \; 1 \; P_4 \; 3 \; P_1 . \qquad (13)$$

We introduce a further item of notation. The pair P_k will be vertically ordered in some box of the current graph; that is, t+k will be on the upper or lower horizontal arc. If t+k is to be placed on the upper arc we denote this by recording P_k , otherwise P_k^*. In either case, of course, it and its companion 3t-k+1 will point to the right. Now rewrite (13) as

$$P_1 \underset{\sim}{7} P_7^* \; 5 \; P_2^* \; 4 \; P_6^* \underset{\sim}{6} P_3 \; 2 \; P_5 \; 1 \; P_4 \; 3 \; P_1 . \qquad (14)$$

Notice that, in the sequence of P's (ignoring subscripts) we change from P to P^* (P^* to P) if and only if the intermediate number is in boldface type. This is called a <u>twist</u>. The sequence (14) is a recipe for filling in figure 12. First it it gives us

$$(15)\qquad \begin{array}{ccccccccccccccc} 8 & & 15 & & 20 & & 16 & & 10 & & 12 & & 11 & & 8 \\ & 7 & & 5 & & 4 & & 6 & & 2 & & 1 & & 3 & \\ 21 & & 14 & & 9 & & 13 & & 19 & & 17 & & 18 & & 21 \end{array}$$

where boldface type is no longer used - it has served its purpose in locating twists.

The currents in the top row will be placed, pointing to the right, at the top of the ladder of figure 12 with 7 rungs; those on the bottom row, on the bottom of the ladder; the currents at the intermediate level on the rungs. The direction of any rung current is uniquely determined by its neighbors and Kcl. The final current graph is displayed in figure 15 and, because of the <u>two</u> twists, is of type $\underset{\sim}{C}$.

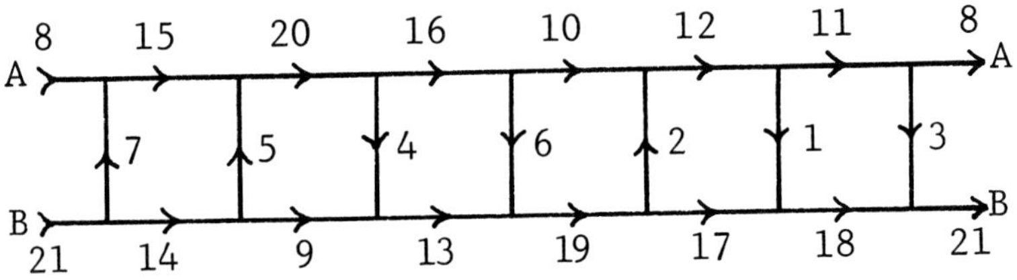

Figure 15

Note that the group is $Z_{43} = Z_{12s+7}$ with $s = 3$ and is the first s for which Heffter's method fails.

The reader is urged to construct the current graph uniquely determined by the zigzag satisfying Z1 to Z4 in figure 16 for $t = 8$.

Once again there are two twists and hence the final current graph will be of type $\underset{\sim}{C}$.

A word should be said to convince the reader that if we obtain (14) from a zigzag satisfying Z1 to Z4 then we can always fill in the current graph of figure 12 so that Kcl holds. If a triple in (14) is of the form $P_a\ c\ P_b$ or $P_a^*\ c\ P_b^*$ then $c = |a-b|$ and the discussion is left to the reader. If we have $P_a\ \underset{\sim}{c}\ P_b^*$ or $P_a^*\ \underset{\sim}{c}\ P_b$, then $c = 2t-a-b+1$, the value of m at $(a+b)/2$. First consider $P_a\ \underset{\sim}{c}\ P_b^*$. At this point in the analogous (15) we get

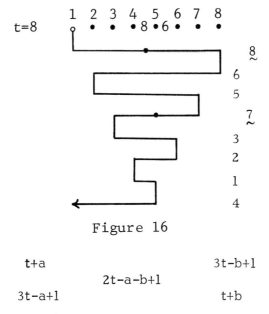

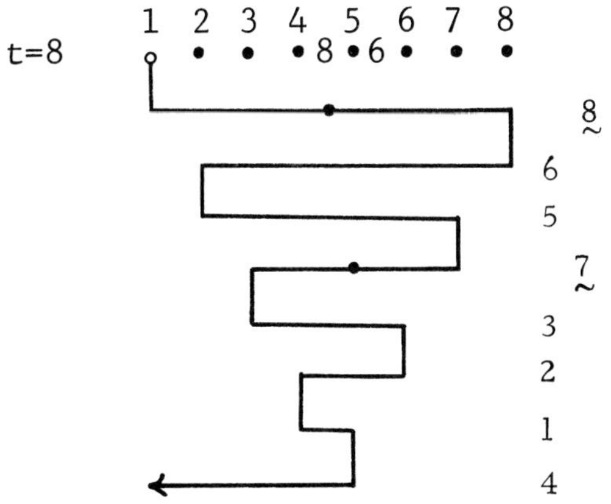

Figure 16

$t+a$		$3t-b+1$
	$2t-a-b+1$	
$3t-a+1$		$t+b$

and the corresponding portion of figure 12 can be filled in with the rung current $2t-a-b+1$ pointing up to conform to Kcl. If $P^*_a \underset{\sim}{c} P_b$ then the display above is reflected about its horizontal axis, and $2t-a-b+1$ points down.

We now propose to record enough zigzags so that the generalization to all t is obvious.

In all cases the zigzag is, geometrically speaking, a simple contracting spiral followed by a final step from $k = [t/2] + 1$ to $k = 1$. The labels however depend upon the residue class of t modulo 4 .

We shall see that if $t \equiv -1$ or $0 \pmod 4$ then the ladder is of type $\underset{\sim}{C}$, since there are two twists; if $t \equiv 1$ or $2 \pmod 4$ then the ladder is of type $\underset{\sim}{M}$ since there is one twist.

If $t \equiv -1 \pmod 4$ then solutions for $t = 3$ and 11 are shown in figures 17 and 18.

The solution for $t = 7$ has already been presented, and the pattern for continuing with $t \equiv -1 \pmod 4$ should be clear.

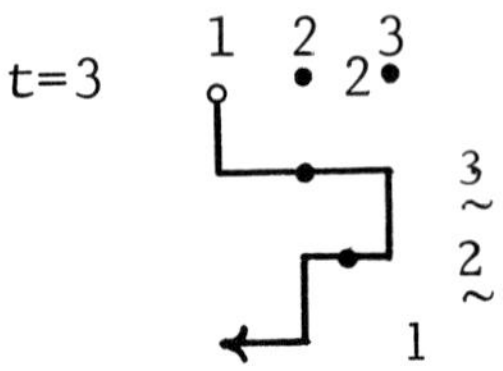

Figure 17

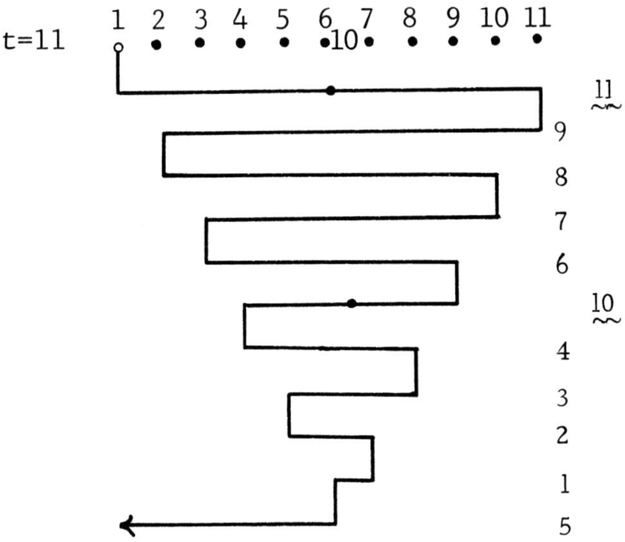

Figure 18

If $t \equiv 0 \pmod 4$ we provide solutions for $t = 4$ and 12 in figures 19 and 20.

The solution for $t = 8$ has already been presented, and again the pattern should be clear.

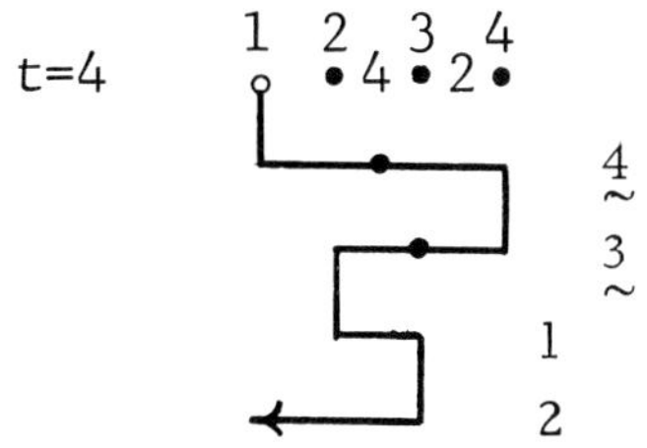

Figure 19

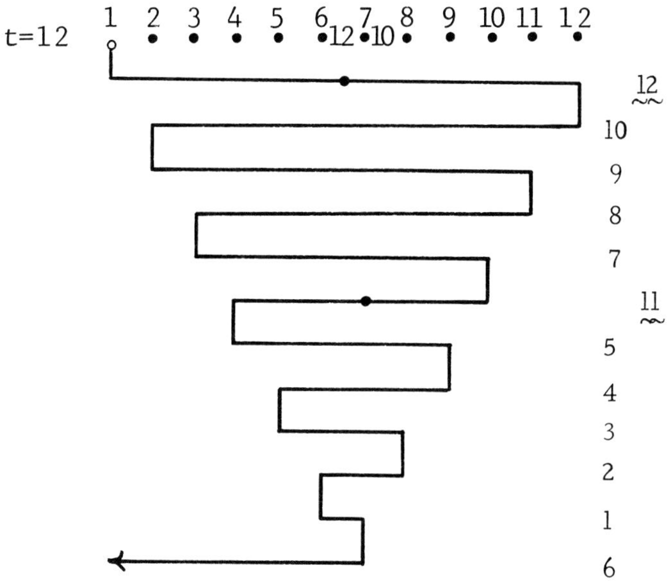

Figure 20

If $t = 1$ we have the current graph of figure 21, Figure 1 shows the determined map $\mathfrak{M}_7$ on a torus. If $t \equiv 1$ (mod 4) the zigzags in figures 22 and 23 are solutions for $t = 5$ and 9 and provide a general pattern.

If $t \equiv 2$ (mod 6) then appropriate zigzags are shown for $t = 2$ and 6 in figure 24 and for $t = 10$ in figure 25.

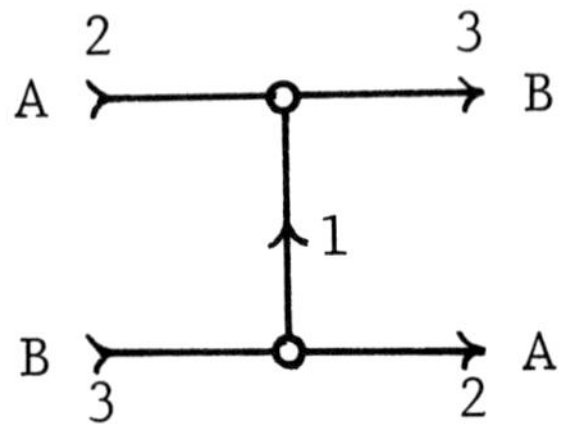

Figure 21

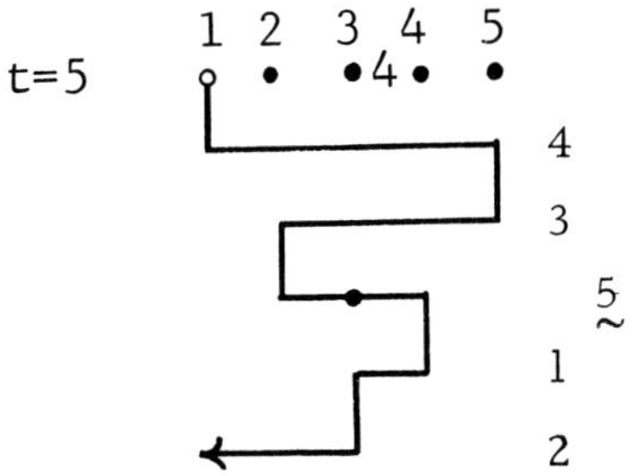

Figure 22

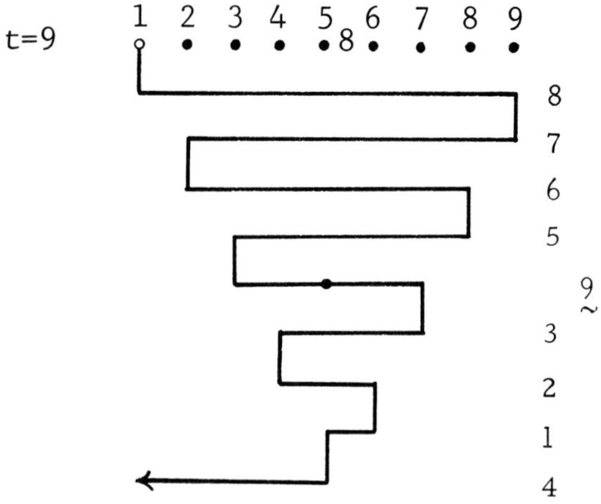

Figure 23

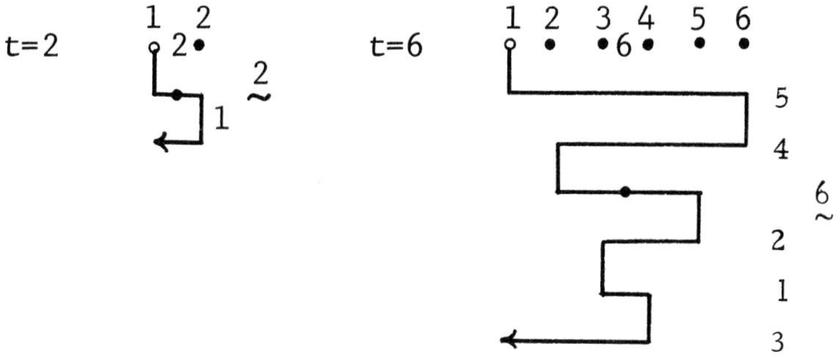

Figure 24

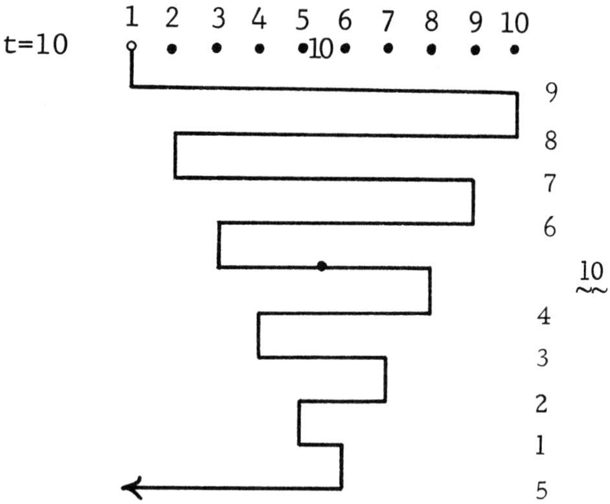

Figure 25

Cases $t \equiv 1$ and $t \equiv 2 \pmod 4$ are really the same. We have spread them out because the weight function used if $t \equiv 1 \pmod 4$ occurs at an integral value of $k = (t+1)/2$ and if $t \equiv 2 \pmod 4$ at a fractional value $(t+1)/2$. The same remarks apply to cases $t \equiv -1$ and $t \equiv 0 \pmod 4$.

At first sight it may appear that, in addition to finding an index 1 orientable scheme for $\mathfrak{M}_n$ with $n = 12s + 7$ (t odd), we have found one for $n = 12s + 1$ (t even). However, though we will have satisfied items 1, 2 and 3 of the theorem, there is a problem with item 4. For each t the ladder will have t rungs. If t is odd, then we can generalize on figure 6 (type $\underset{\sim}{C}$) or figure 7 (type $\underset{\sim}{M}$) and place rotations on the vertices so that we get a single circuit. If t is even, then for <u>no</u> ladder of type $\underset{\sim}{C}$ or $\underset{\sim}{M}$ is this true. Consequently with the application to the Heawood conjecture here under consideration, the current graphs for even t must be considered combinatorial oddities - they are useful, however, in another context.

<u>This completes our objective of providing a Heffter type solution for the Heawood conjecture in case 7 .</u> However, a slight inelegance remains. If $t \equiv -1$ <u>or</u> $0 \pmod 4$ we get a $\underset{\sim}{C}$ type ladder and if $t \equiv 1$ or $2 \pmod 4$ the ladder is of type $\underset{\sim}{M}$.

It is easy to show that if $t = 1$ or 2 , then the ladder <u>must</u> be of type $\underset{\sim}{M}$. For all other values of t we shall display an $\underset{\sim}{M}$ type solution if $t \equiv -1$ or $0 \pmod 4$ and a $\underset{\sim}{C}$ type solution if $t \equiv 1$ or $2 \pmod 4$.

To do this we propose to construct a current graph $(K_t, Z_{6t+1}, \lambda_t)$ where K_t is a ladder with t rungs. We require (11) 1) and 2) to hold but change 3) to

(16) $1, 2, 3, \ldots t-1, \hat{t}, t+1$ appear on the rungs.

The remaining currents will be considered in pairs P_k as before, but now the admissible values of k are $0, \hat{1}, 2 \ldots t$. The omission of P_1 means that we do <u>not</u> propose to use $t+1, 3t$ as a pair of horizontal currents. However, $t+1$ (but not t) is to be used as a rung current, and $-3t = 3t+1$ in Z_{6t+1} will be used <u>with</u> t as pair P_0 . The currents on the graph will therefore be

(17) $1 \ 2 \ 3 \ \ldots . \ 3t-1 \ \widehat{3t} \ 3t+1 .$

The set $\pm k$ for k in (17) exhaust $Z_{6t+1} \setminus 0$ without repetition, hence, if all this can be done, λ_s will be 1-1 and onto.

The basic data in dot form is displayed in figure 26 where 1 must be avoided and so is crossed out.

0	1	2	3		t-2		t-1		t
o	x	•		... 6	•	4	•	2	•

Figure 26

The zigzag will have the properties listed below and consequently provide a solution. Note, however, in reference to figure 12, that the currents at the left will be t and $3t+1$ instead of $t+1$ and $3t$.

Z1. The zigzag starts and terminates at $k = 0$ with one riser under each $k = 2, \ldots, t$.

Z2. The labels, recorded to the right of the successive steps, are the collection of numbers in (16).

Z3 and Z4 remain unchanged.

The solution for $t=3$ does not follow the general pattern and is displayed in figure 27. From the solutions for $t = 7$ and 11 in figures 28 and 29 the general pattern should be clear for $t \equiv -1 \pmod 4$.

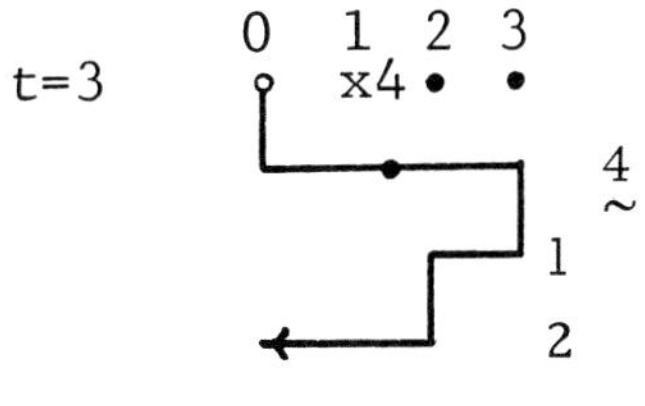

Figure 27

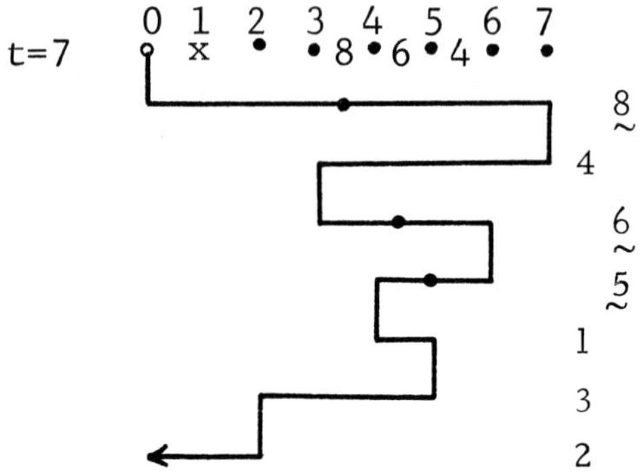

Figure 28

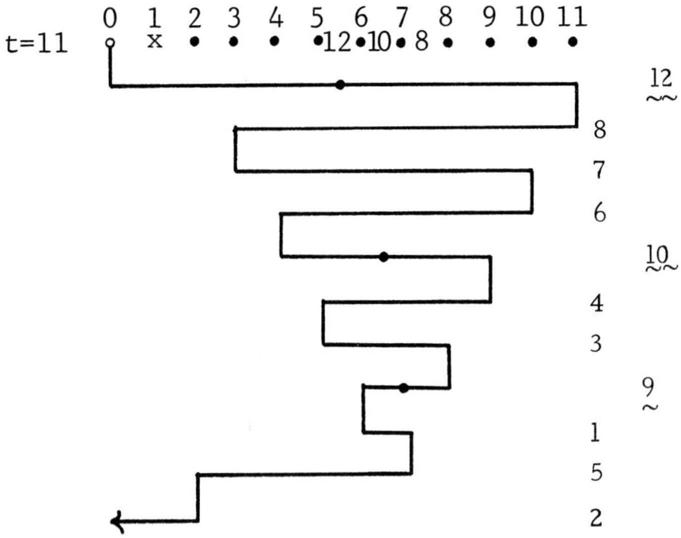

Figure 29

If $t = 4$ the pattern is exceptional and displayed in figure 30.

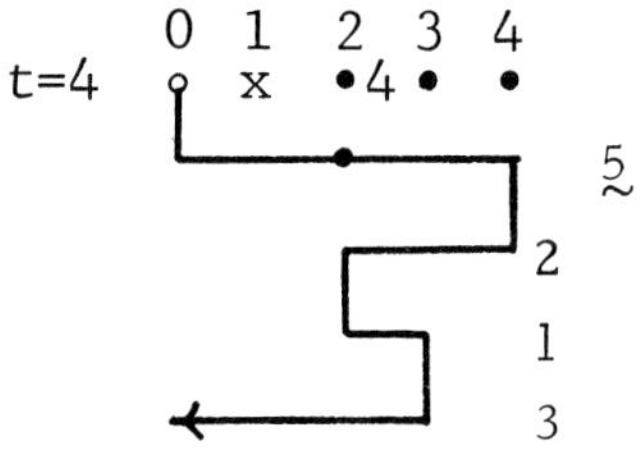

Figure 30

From the solution for $t = 8$ and 12 in figures 31 and 32 the final pattern should be clear for $t \equiv 0 \pmod 4$.

If $t \equiv -1$ or $0 \pmod 4$, then there are an odd number (1 or 3) of twists and so the ladder must be of type $\underset{\sim}{M}$.

If $t = 5$ and 9, then figures 33 and 34 shows easily generalized solutions for $t \equiv 1 \pmod 4$.

If $t = 6$ and 10 then figures 35 and 36 shows the pattern for $t \equiv 2 \pmod 4$.

If $t \equiv 1$ or $2 \pmod 4$, then there are two twists and therefore the ladder is of type $\underset{\sim}{C}$.

V. Concluding remarks.

It is hoped that the contention of the introduction, that the mystery of Heffter's inability to solve all of case 7 was due to his lack of knowledge of current graphs and rotations is convincing.

In fact, had Heffter had these tools available he would no doubt have discovered much simpler solutions to case 7. (See below and [11].)

As a concluding remark, Ringel [6], who first solved case 7 completely, did it in two sub-cases depending on the parity of s. Had he known these techniques he would certainly have achieved the result in a single case. The only difference between his two cases is a rotation at one vertex. If s is even, then, in our terminology, he

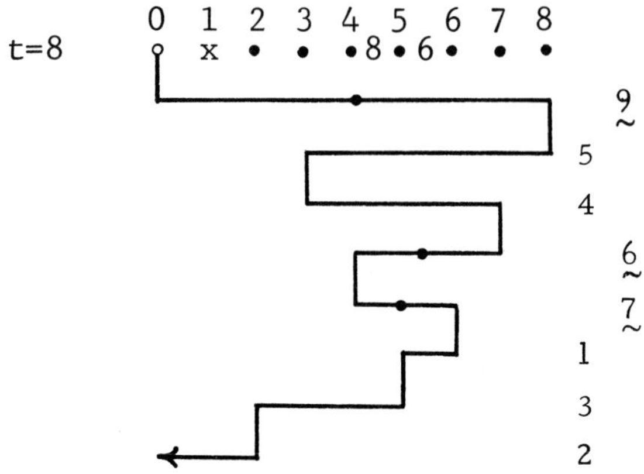

Figure 31

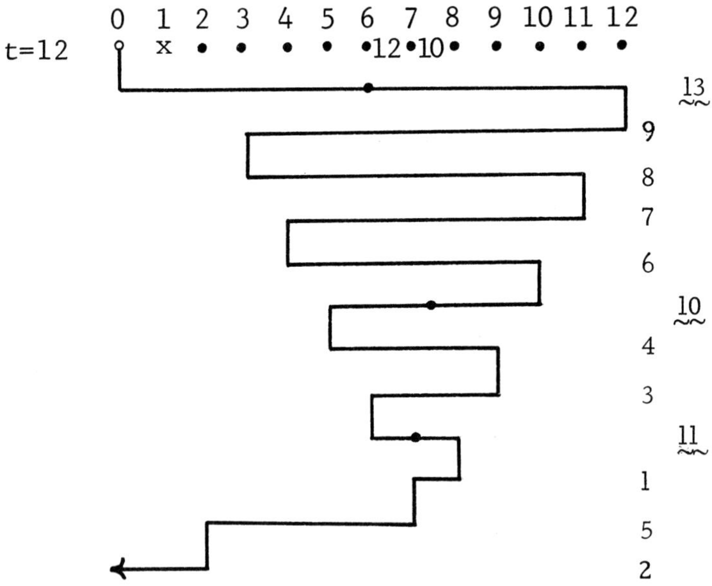

Figure 32

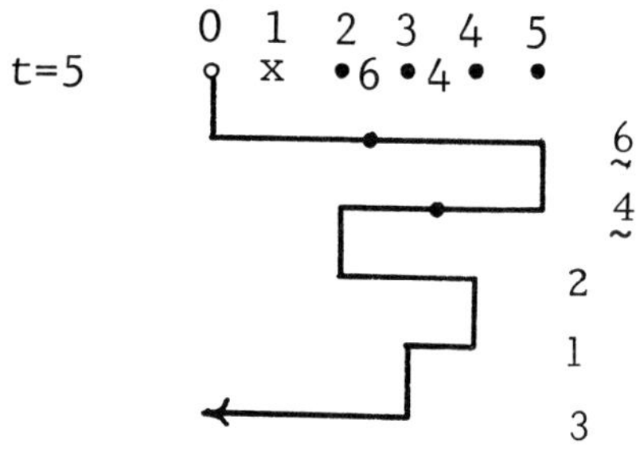

Figure 33

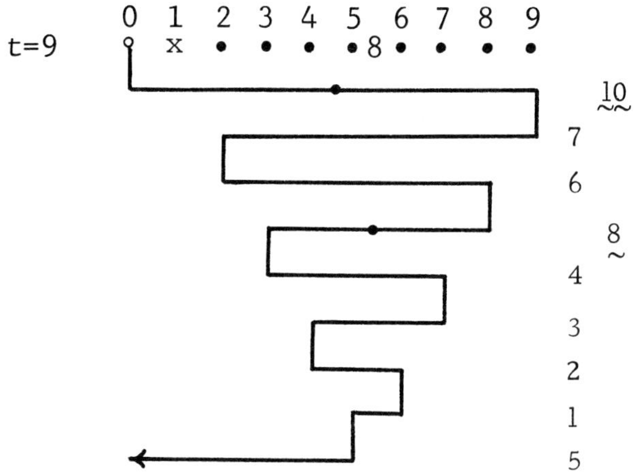

Figure 34

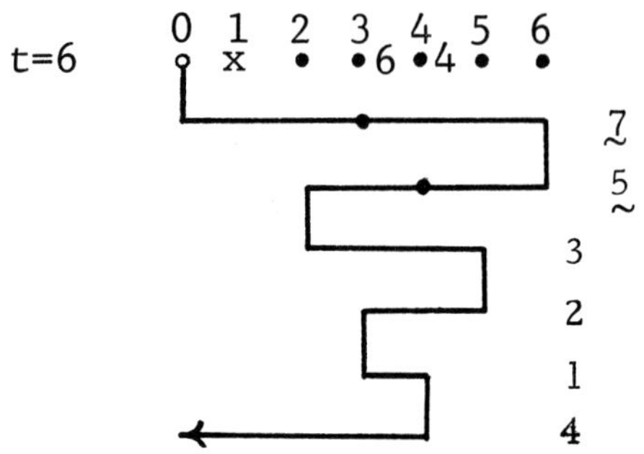

Figure 35

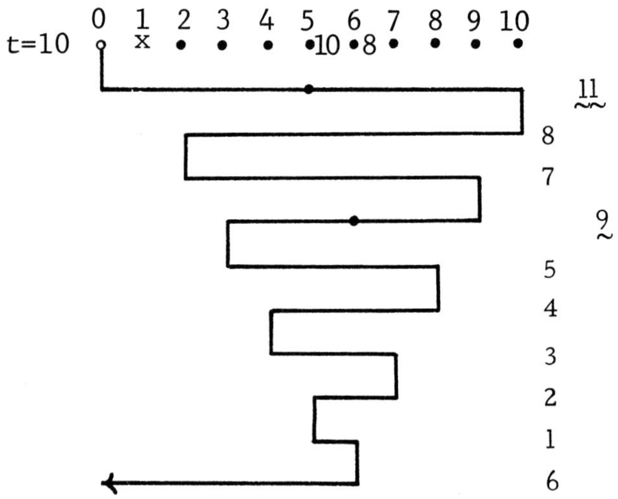

Figure 36

generalized the rotations following the pattern of figure 8 , if s is odd that of figure 9 . The only difference, from the point of view of the rotation, between these figures is the rotation at the first vertex, but it makes an enormous difference in the derived row 0 . On the other hand either of these patterns of rotation can be generalized, so the matter can be handled in a single case.

REFERENCES

1. Richard Courant and Herbert Robbins, What is mathematics? Oxford University Press, London, 1941.

2. W. Gustin, Orientable embeddings of Cayley Graphs, Bull. Amer. Math. Soc. 69 (1963), 272-275.

3. P. J. Heawood, Map colour theorem, Quart. J. Math. 24 (1890), 332-338.

4. L. Heffter, Über das Problem der Nachbargebiete, Math. Ann. 38 (1891), 477-508.

5. G. Ringel, Färbungsprobleme auf Flächen und Graphen, VEB Deutscher Verlag der Wissenschaften, Berlin, 1959.

6. G. Ringel, Über das Problem der Nachbargebiete auf orientierbaren Flächen, Abh. Math. Sem. Univ. Hamburg 25 (1961), 105-127.

7. G. Ringel and J. W. T. Youngs, Solution of the Heawood map-coloring problem , Proc. Nat. Acad. Sci. U. S. A. 60 (1968), 438-445.

8. G. Ringel and J. W. T. Youngs, Solution of the Heawood map-coloring problem-case II , J. Combinatorial Theory 7 (1969), 71 - 93.

9. G. Ringel and J. W. T. Youngs, Remarks on the Heawood Conjecture, Proof Techniques in Graph Theory, Academic Press, London and New York, 133-138.

10. J. W. T. Youngs, The Heawood map-coloring conjecture, Graph Theory and Theoretical Physics, Academic Press, London and New York, 313-354.

11. J. W. T. Youngs, The Heawood map-coloring problem-cases 1, 7 and 10, J. Combinatorial Theory, (to appear).

This research reported on in this article was supported in part by the National Science Foundation.

Graph Theory Algorithms

RONALD C. READ

1. Introduction

In this paper I should like to discuss some algorithms which are of use in graph theory -- algorithms, which, for example, enable us to determine whether a given graph has such and such a given property, or to construct graphs or subgraphs of a particular kind. There are many such algorithms, and their number grows daily. Accordingly, rather than attempt to cover the whole field, I shall deal with only a small, and probably quite unrepresentative selection, containing, hopefully, some of the more interesting examples.

The subject of graph-theoretical algorithms is as old as graph theory itself, which is usually regarded as having started with Euler's famous solution of the problem of the seven bridges of Königsberg. In generalizing this problem, Euler examined the possibility of traversing a connected graph (or, more strictly, a multigraph, since we can allow multiple edges here) by a walk starting and ending at any node, and including, exactly once, every edge of the graph. He showed that this is possible if and only if every node of the graph has even valency. Such a graph is now called an Euler graph, and we can call such a walk an "Euler walk". In general, an Euler graph will have many different Euler walks.

I have not read Euler's original paper [2] on this subject (I wonder how many people have!) but I believe that he proved the sufficiency part of the above theorem by giving an algorithm for constructing an Euler walk. It is very simple and will serve as an easy example with which to begin our excursion into the realm of graph-theoretical algorithms.

Let G be a finite connected Euler graph. Informally stated the algorithm says "Start anywhere and go anywhere until you can't go any further. Start again with what is left of the graph, and continue like this until there isn't any left. Then join up the pieces." More formally, it can be stated as follows:

Step 1. Starting at any node construct a walk, W, eventually finishing at the starting node.

Comment: Since the valencies are even, whenever we reach a node other than the starting node there is always an edge by which we can leave it. We are therefore forced to stop the walk only when we arrive at the starting node and find that all the edges incident with it are now part of the walk. (See figure 1, with node A as starting node).

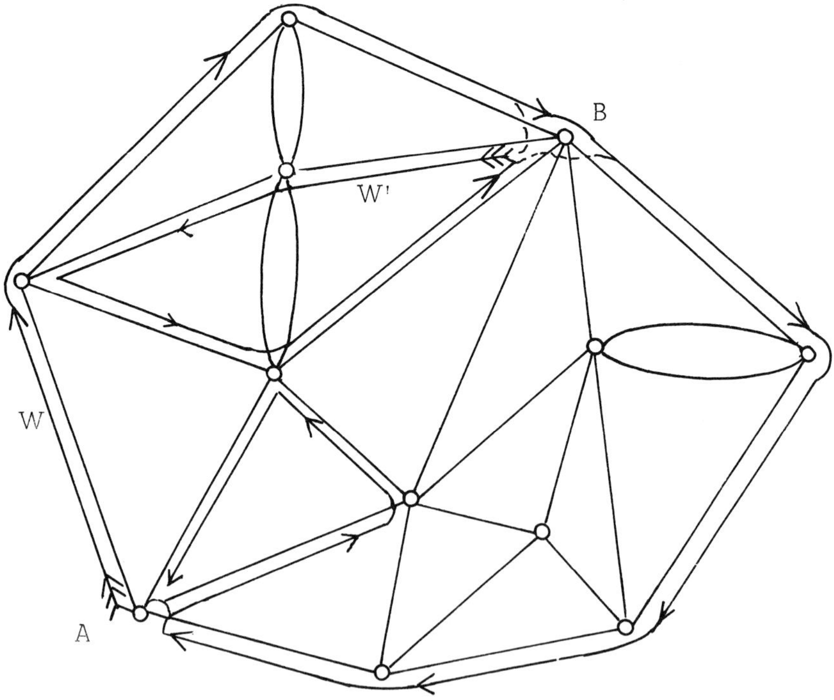

Figure 1

Step 2. If W contains all edges of G, then we are finished. If not, then G - W is an Euler graph, and has at least one node, B, say, in common with W (otherwise G would not be connected.)

Step 3. As in Step 1, construct a walk, W', in G - W, beginning and ending at B.

Step 4. Combine walks W and W' as follows: Traverse W until node B is reached; then traverse W', back to B; then complete the rest of walk W.

Step 5. Regard this combined walk as a new walk W in G, and go back to Step 2.

The algorithm must terminate since the number of edges in the walk W increases, no edge is included more than once, and G is finite.

2. Spanning trees.

Another algorithm from the early (i. e. pre-twentieth century) history of graph-theory is that for constructing a spanning tree of a connected graph. This was first devised by Kirchhoff [8] in his treatment of electrical circuit theory. A spanning tree T of a graph G is a subgraph of G which (a) is a tree, and (b) contains all the nodes of G. To construct such a spanning tree we keep choosing edges of G, one at a time, subject to the proviso that we do not choose an edge which, when taken with some of those already chosen, completes a circuit in G. More formally, we progressively construct sets $\mathfrak{y}$ and $\mathfrak{n}$ of edges in the following way:

Step 1. Take $\mathfrak{y}$ and $\mathfrak{n}$ to be empty.

Step 2. Choose any edge, a, of G which is not in $\mathfrak{y} \cup \mathfrak{n}$.

Step 3. Look at the subgraph H determined by the edges $\{a\} \cup \mathfrak{y}$.

Step 4. If this subgraph has a circuit, include a in the set $\mathfrak{n}$. Otherwise include it in $\mathfrak{y}$.

Step 5. If $\mathcal{Y}$ now contains p - 1 edges, where p is the number of nodes in G, the algorithm terminates, and the edges in $\mathcal{Y}$ form the required spanning tree. Otherwise go back to Step 2.

The importance of this construction in electrical circuit theory is easy to see. Suppose we have an electrical network shown diagramatically in figure 2 as a graph, where the edges represent series combinations of resistors and sources of e. m. f. The distribution of currents in the network is determined by equations that can be written down according to the structure of the network. Each node gives an equation (Kirchhoff's first law) and every circuit gives an equation (Kirchhoff's second law). But the latter equations are not independent. Thus in figure 2 the equation derived from the circuit aefghd will be a linear combination of those derived from the circuits abcd and befghc.

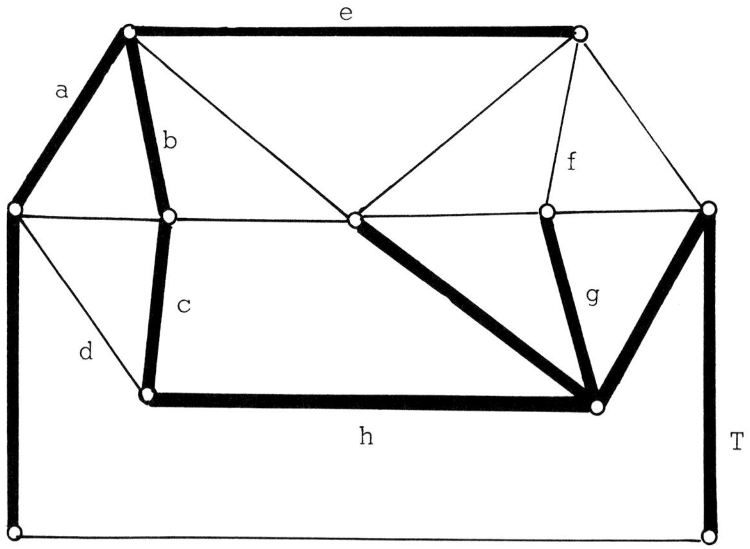

Figure 2

Let us make this more precise. If we label the edges of a graph G with the integers 1, 2, 3, ..., q, say, we can specify any subgraph H of G by a q-dimensional binary vector

$$\underline{x} = (x_1, x_2, x_3, \ldots., x_q)$$

where $x_i = 1$ if edge i is in H, and $x_i = 0$ if it is not. Suppose that two circuits in a network are given by vectors $\underline{x}$ and $\underline{y}$. Then the subgraph $\underline{x} \oplus \underline{y}$, where "$\oplus$" denotes element-wise modulo 2 addition, may or may not be a circuit; but if it is, then the equation to which it gives rise will be dependent on those given by $\underline{x}$ and $\underline{y}$.

Thus it becomes important to form the Kirchhoff equations only for a set of circuits that are known to be independent in this sense, and this is easily done once a spanning tree of the graph has been constructed. In figure 2 a spanning tree T is shown in heavy lines. If we add to T any edge not in T we automatically complete one circuit (the addition of edge d in figure 2 completes the circuit abcd). By adding to T, in turn, each of the $q - p + 1$ edges of G - T, we obtain $q - p + 1$ circuits, and these will be independent. This explains the interest of Kirchhoff, and electrical engineers in general, in spanning trees.

This interest does not stop there, however. For some purposes the electrical engineer is not content to find one spanning tree of a given network--he wants to find them all. Now this is a much more tricky problem and not so easily solved. It is not difficult to see how, given one spanning tree, we can find another. We have already seen that the addition of a non-tree edge to a spanning tree completes a circuit. If we remove another edge of this circuit we obtain a new spanning tree, and we can thus go from one spanning tree to another. It is also quite easy to show that we can go, in this way, from any given spanning tree to any other. What is not so easy is to do this systematically, that is, in such a way that (a) every spanning tree of G is reached at some stage in the process, and (b) no spanning tree is reached more than once. Requirement (a) is essential;

requirement (b) is highly desirable, especially if the graph is a large one. It turns out that this can be done and a description of such a method is given by Mayeda and Seshu [13].

3. Circuits.

We have seen that from a spanning tree we can obtain a set of independent circuits (a fundamental set). Suppose now that we want to find all the circuits in a given graph.

We could do this by constructing all possible combinations of fundamental circuits (i. e. circuits in the fundamental set), and combining these in the sense of "modulo 2 addition" as given above. Since there are $q - p + 1$ non-tree edges, there are $2^{q-p+1}-1$ possible combinations, excluding the empty one. They can be formed systematically as follows: we number the fundamental circuits $\underline{x}_1, \underline{x}_2, \underline{x}_3, \ldots$ in some order, and form the set $\{\underline{x}_1, \underline{x}_2, \underline{x}_1 \oplus \underline{x}_2\}$; then we form all combinations of $\underline{x}_3$ with the elements of this set, and so on. Unfortunately not all such combinations will be circuits. In figure 3 the edges a, b and c give rise to circuits $\underline{x}$, $\underline{y}$ and $\underline{z}$ when added to the spanning tree indicated. The combinations $\underline{x} \oplus \underline{y}$, $\underline{y} \oplus \underline{z}$ and $\underline{z} \oplus \underline{x}$ are all circuits, as can be seen; but $\underline{x} \oplus \underline{y} \oplus \underline{z}$ is not, consisting, as it does, of two separate circuits. Thus many of the combinations of fundamental circuits will have to be discarded.

If we were to discard such an inadmissible combination of circuits as soon as it was formed during the systematic construction of combinations described above, we should run the risk of missing something; for it could well happen that one of the circuits we wanted was a combination of one of these inadmissible combinations with a circuit, or even with another inadmissible combination. On the other hand, to form all $2^{q-p+1}-1$ combinations, and then throw out those that are inadmissible is exactly the sort of thing one tries to avoid in an efficient algorithm. It has been shown by Welch [18], however, that it is not necessary to do this. Provided that we order the fundamental circuits $\underline{x}_1, \underline{x}_2, \underline{x}_3, \ldots$ according to a certain set of rules, we can cheerfully throw out any inadmissible combinations as soon as they appear,

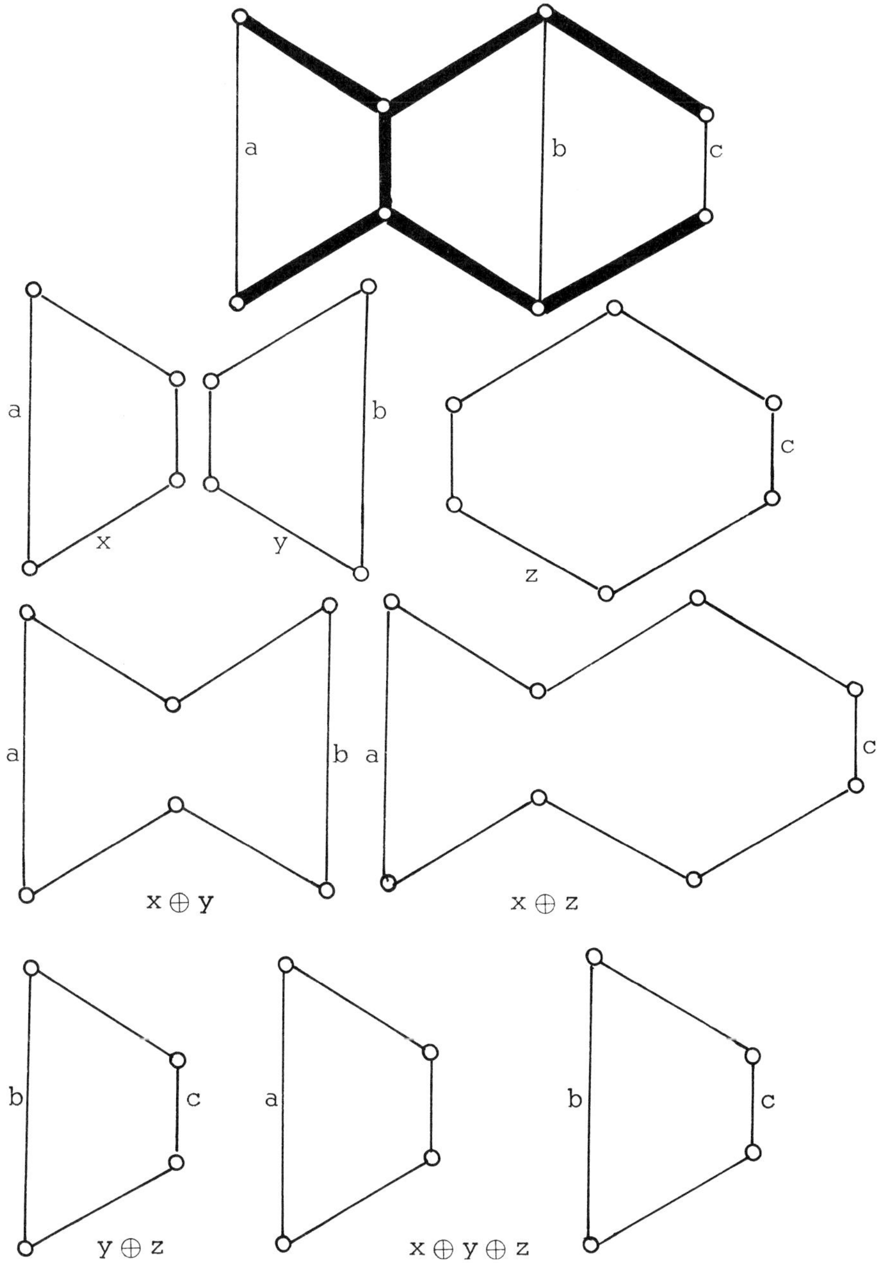

Figure 3

without fear of missing anything, and thus keep to a more manageable size the set of circuits with which the next fundamental circuit in the list is to be combined.

4. Bridges, blocks, and cutnodes.

The usefulness of the spanning tree algorithm is by no means confined to the applications given so far. Let us look at some more. We can first remark, almost in passing, that the algorithm immediately gives us a test for connectedness. If we apply it to a graph G which may or may not be connected, then either we obtain a spanning tree -- in which case G is clearly connected; or we run out of edges before the set $\mathcal{Y}$ has achieved its full complement of p-1 edges, in which case G is not connected.

Suppose we want to know whether a graph G has any "bridges", that is, edges whose removal disconnects the graph (such as the edge PQ in figure 4). A bridge is characterized by belonging to no circuit, and hence no fundamental circuit; whereas any edge that is not a bridge belongs to at least one fundamental circuit. Hence if we construct a spanning tree of G, form each fundamental circuit in turn, and note the edges which comprise it, then at the end of the operation any edge that has not been noted at any stage will be a bridge. All "non-bridges" will have been noted as belonging to some fundamental circuit.

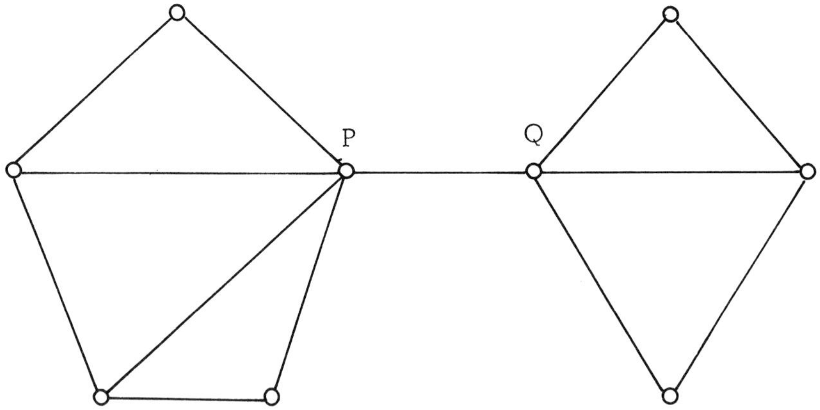

Figure 4

If this process is to be carried out on a computer, the only stage which may cause some bother is that of distinguishing the edges of a fundamental circuit, formed by the addition of a non-tree edge, from the other edges that may be hanging around at the time. Elsewhere [15] I have given an easy method for doing this and shown also that a great deal of the business of checking the edges that are not bridges can go on while the spanning tree is in process of construction.

We can elaborate this still further. Let us attempt to identify the blocks and cutnodes of a graph, where a cutnode is a node whose removal disconnects the graph, and a block is a maximal subgraph having no cutnodes. Figure 5 shows a graph having one cutnode C, which (in an obvious sense) divides the graph into three blocks. If we can derive

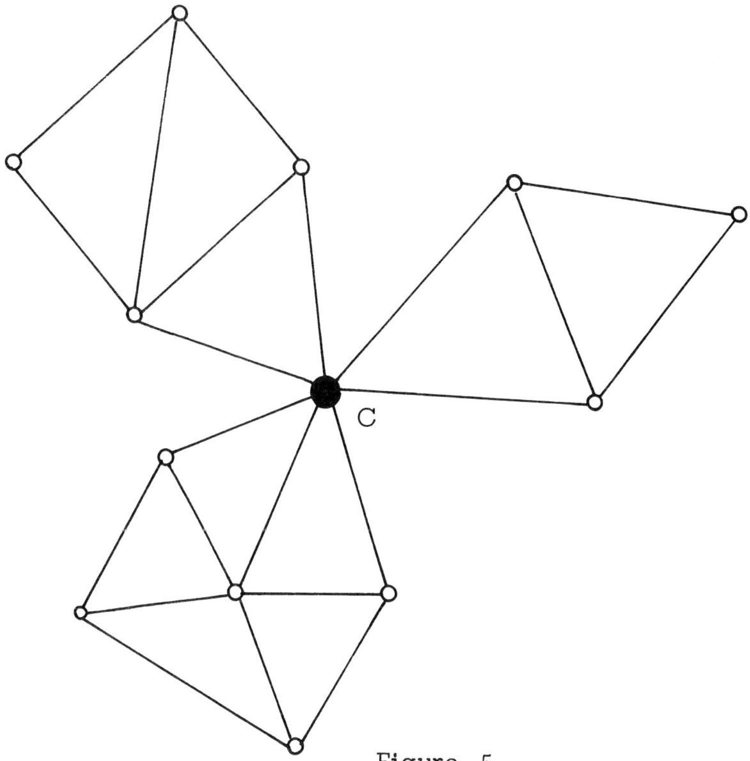

Figure 5

an algorithm, which for any pair of edges will determine whether they are in the same or different blocks, then we can identify the blocks of G. This done, we can identify the cutnodes; they will be those nodes, and only those, which belong to more than one block.

Clearly, two edges in different blocks cannot occur in the same fundamental circuit. If two edges are in the same block then there will be at least one circuit which contains both of them; but it may not be a fundamental circuit, a fact which causes a slight complication. Construct a fundamental set of circuits in the graph G, and for each fundamental circuit, record as "equivalent" the edges which comprise it. The slight complication just mentioned is that two edges in the same block, but not belonging to a common fundamental circuit will not, by this procedure, be recorded as "equivalent". We get over this difficulty by a transitivity rule, viz, that two edges that are "equivalent" to a third are "equivalent" to each other. The relation that we have set up between the edges then becomes an equivalence relation, and we can drop the quotation marks that have appeared in this paragraph. Any bridges that the graph may have will have been found in passing, as above. These are blocks in themselves, and the equivalence classes just constructed will determine all the other blocks of the graph. The identification of the cutnodes follows easily, as already remarked.

5. Cut-price railroad networks.

We have not yet left the topic of spanning trees--a topic which crops up again and again in the discussion of graph-theoretical algorithms--but we now look at a different kind of application, and consider graphs in which each edge has a number associated with it. Such graphs are called networks, a term we have already used. Frequently one defines a function over all subnetworks of a network, and then asks for a subnetwork, satisfying certain conditions, for which this function is a maximum or a minimum. The following problem is of this type.

A small country, having no railroad system, wishes to construct a railroad network which will link all its principal towns. Junctions where three or more railroad tracks meet must occur only at the towns, it must be possible to travel from any of the towns to any other, and the network must be constructed at the minimum cost. This sounds like a linear programming problem, and can be formulated as such; but the graph-theoretical approach is more interesting.

What we have here is a network, whose nodes are the towns. Between any two nodes there is an edge (a possible railroad track), and with each edge there is associated a number -- the cost of constructing that track. We require a subnetwork (a) which is connected and which spans the network (since every town must be accessible from every other) and (b) for which the total cost of the edges is a minimum. It is clear that the required subnetwork will be a tree, for if it contained a circuit one edge of the circuit could be removed with a consequent diminution of the total cost. The problem is thus that of finding a minimum spanning tree of a given network. We can assume that cost is proportional to length of track (i. e. length of the edges of the network).

An algorithm for doing this has been given by Kruskal [11] and is very simple. It is identical to the algorithm given above for constructing any spanning tree, except that at each stage we consider for inclusion in the set $\mathfrak{y}$, the shortest of the edges not belonging to $\mathfrak{y} \cup \mathfrak{n}$ (or one of these, if there is a tie for the shortest edge). Kruskal assumed that the edges were all of different lengths (in which case the minimum spanning tree is unique). The following modification of his proof shows that, if we do not make this assumption, the algorithm will yield <u>one</u> of the several possible minimal trees.

Let $a_1, a_2, a_3, \ldots, a_{p-1}$ be the sequence of edges chosen by the algorithm and let T be the tree that they determine. Let T_1 be any minimal spanning tree and let a_i be the first edge in the sequence which is not in T_1. The addition of a_i to T_1 creates a circuit C. No edge of C can be longer than a_i, for then the replacement of this edge by a_i would give a tree of shorter total length than T_1. On the other hand, the edges of C cannot all be shorter than a_i, since they would then all belong to T, and this would give T a circuit. Hence at least one edge of C is of the same length as a_i. By removing this edge and adding a_i we get a tree T_2 of the same total length, and hence also a minimal tree, which agrees with T up to and including a_i. Repeating this process we eventually find a minimal tree which agrees with T up to and including a_{p-1}, which proves the theorem.

Other algorithms (such as that of Dijkstra [1]) have been given for solving this problem, and have some advantages over the above. Kruskal's algorithm is probably the most straightforward.

Note that if we do not stipulate that junctions occur only at the towns, but allow them to be sited anywhere, then we have a completely different problem. For example, if there are four towns at the corners of a unit square, the minimum tree in the original problem has length 3 (figure 6a); whereas without the restriction on the siting of the junctions there is a minimum tree of length $\sqrt{3}+1$ (figure 6b). The latter problem is very much more difficult, and as far as I know, very little progress has been made with it.

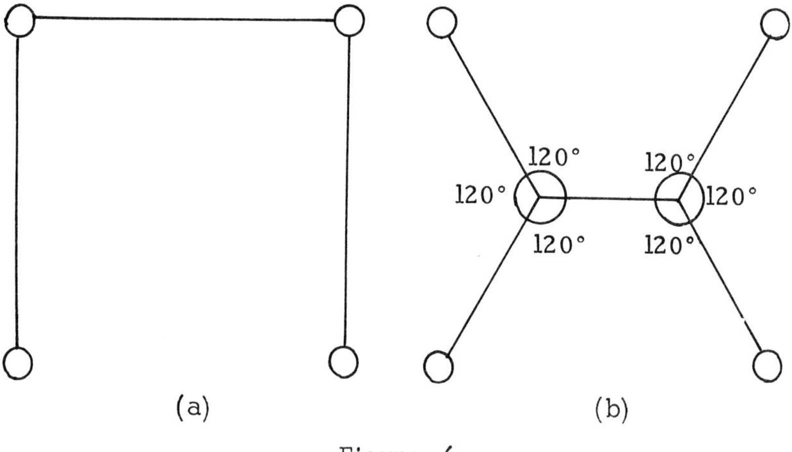

Figure 6

6. Planarity

An interesting algorithmic problem is that of determining whether a given graph is planar or non-planar. A planar graph is one that can be drawn in a plane with no edges crossing. There is a well-known criterion for a graph to be planar, which is due to Kuratowski [12], and says that a graph is planar if and only if it does not contain as a subgraph a homeomorph of one of the two "Kuratowski" graphs, viz., the complete graph K_5 and the complete bipartite graph $K_{3,3}$. A homeomorph of a graph is one that can be obtained from it by inserting nodes of valency 2 into its edges.

This criterion is very pretty, but it does not provide any very practical algorithm for determining the planarity or non-planarity of a given graph. To do this and to construct a planar drawing of the graph if it turns out to be planar, requires a more indirect approach to the problem.

We shall first consider an algorithm described by Weinberg [17]. We start by constructing a spanning tree of the given graph G, and we draw it in the plane in any convenient manner. We now try to add the remaining edges so that no crossings occur. This may not be possible; for even if G is planar, we may have drawn the tree in a position which it cannot have in any plane drawing of G, and we shall reach a stage where none of the remaining edges can be added. This will be so, for example, with the partially drawn graph of figure 7 if we are required to add edges from P_3 to P_7, and from P_2 to P_4 and P_9. What has happened is that in adding the edges to the spanning tree, we have divided the plane into regions, and we now find that the two nodes to be joined are not on the boundary of a common region. But all is not lost; the graph may still have a certain amount of mobility which will enable us to twist it around so as to make possible the desired connexion.

For instance, if the graph so far constructed has a cutnode, any block incident with that cutnode can be swivelled around it into a variety of positions. In figure 7 P_1 is a cutnode, and the edge P_1P_2 can be swivelled about P_1 so that P_2 lies in the region A, as in figure 8. Again, if the graph is 2-connected, i. e., has a pair of nodes whose removal breaks the graph into portions (as with P_5 and P_8 in figure 7) then these two portions can be independently rotated about the two nodes. Thus by rotating the right-hand portion of figure 7, containing nodes P_5, P_6, P_7 and P_8, we bring it to the position shown in figure 8. In figure 8 the required

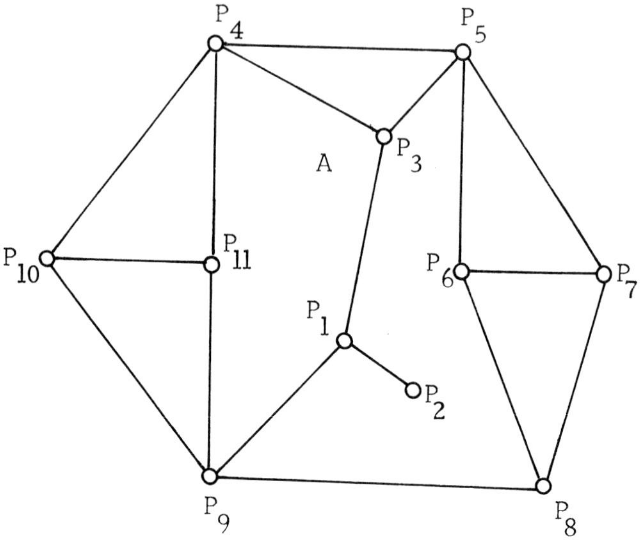

Figure 7

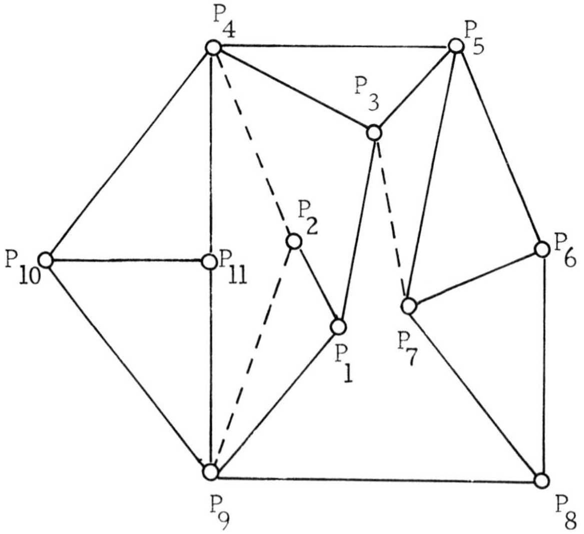

Figure 8

edges can now be added, as has been done.

It is known that a planar 3-connected graph (one in which at least 3 nodes must be removed to disconnect it) has essentially only one plane imbedding. Hence, once we obtain a 3-connected graph, any subsequent failure to insert another edge will show the graph is non-planar. On the other hand there are many non-planar 2-connected graphs, so that we may find that no amount of swivelling will enable us to accomodate an edge which has to be added.

It is not easy, on a computer, to detect pairs of nodes whose removal disconnects the graph, and in the early stages of construction the graph may have many degrees of freedom, which make for tricky programming. Thus this method is not too well adapted for computer implementation. It is very convenient when carried out by hand, however, since the eye is quick to perceive that a graph is 2-connected, and to see what sort of swivellings and twistings will enable new edges to be inserted.

A different approach to the problem of testing for planarity was made by Tutte [16], and Fisher and Wing [4], who took as their starting point a circuit rather than a tree. The method is as follows.

Take any circuit of the graph G, such as that shown in heavy lines in figure 9. If the graph is planar, then in any drawing of it in the plane, this circuit can be deformed into a circle Σ, and if we remove from the graph all the edges of Σ, together with all the other edges of G that are incident with nodes of Σ (we shall call these latter edges "link" edges), then what is left will be a number of connected components (enclosed in dotted circles in figure 9) each of which will be either inside or outside the circle Σ. We need a name for these components, for we shall get into difficulties if we continue to call them components. Let us call them "fragments", and denote them by F_1, F_2, F_3, ... Each F_i is connected to Σ by a certain set L_i of link edges. At this stage we must also take into account any "transversals" that there may be, i. e. edges joining one node of Σ to another, such as the edge a in figure 9. The easiest way of handling these transversals is to bring them in line with the

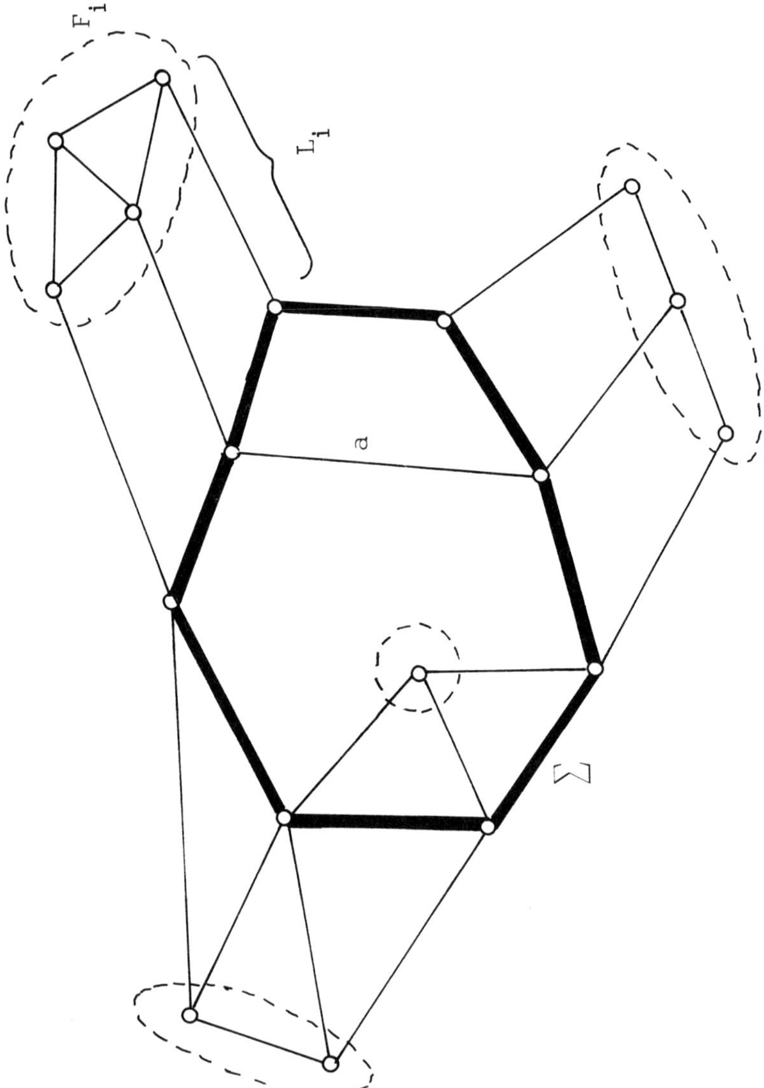

Figure 9

fragments F_i by inserting a node in the middle of each. Each transversal then becomes a fragment (consisting of one node) joined to Σ by two link edges. We can include these fragments along with the others. Let H_i denote the subgraph of G determined by Σ, F_i and L_i (an example is shown in figure 10). Suppose now that we can show that each H_i is planar. Then we can derive a plane drawing of G, provided we can determine for each F_i, whether it is to go inside or outside Σ, and do this in such a way that no link edges will intersect. Let us look first at this latter requirement.

Since we are not, at this stage, interested in the detailed structure of the F_i we can reduce them to points as in figure 11. In this figure we have an example of two "incompatible" fragments, F_1 and F_2; we cannot draw them both on the inside (or both on the outside) of Σ without causing some link edges to cross. If two fragments F_i and F_j are incompatible then one must go inside Σ, and the other outside.

We therefore see whether we can place some fragments inside Σ and the rest outside in such a way that all pairs of incompatible fragments are differently placed. If we cannot do this, then G is not planar; if we can, then we are left with the task of determining whether the several graphs H_i are themselves planar or not. The problem is thus reduced to several similar problems with smaller graphs, and by applying the same procedure to each of these, we eventually obtain a verdict one way or the other.

This account of the algorithm is slightly oversimplified --there are a few complications that can arise, but they need not concern us at this stage.

It is not difficult to tell from the way their links are connected to the nodes of Σ whether two fragments are incompatible or not. Fisher and Wing suggest the formation of a matrix $[a_{ij}]$, where $a_{ij} = 0$ if F_i and F_j are compatible, and $a_{ij} = 1$ if they are not. From such a matrix one can tell whether a suitable allocation to inside and outside is possible. When we started to program this algorithm for our computer, it seemed to us a pity to form the whole matrix before examining it. It could well happen that the first three fragments

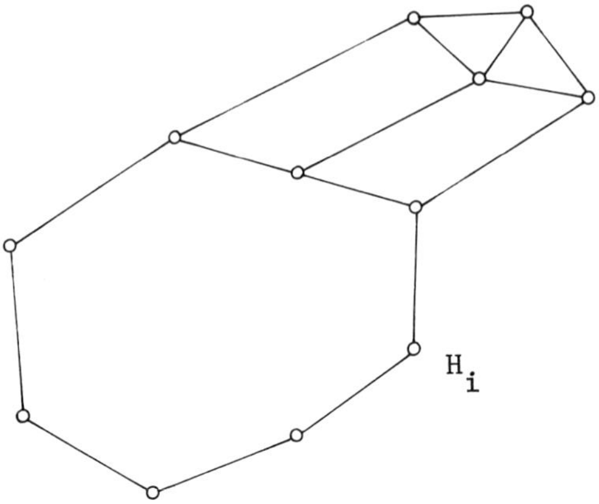

Figure 10

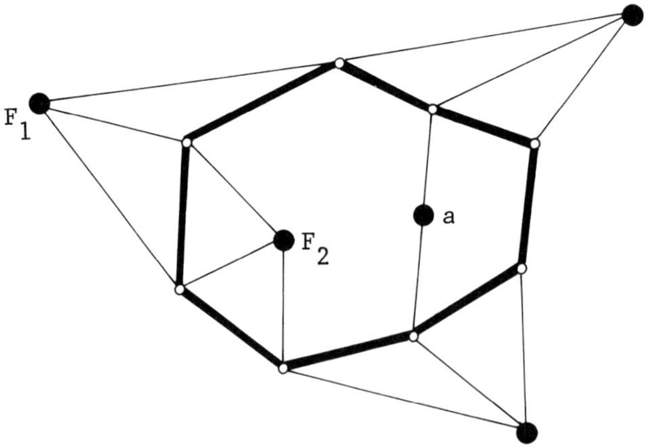

Figure 11

to be looked at are mutually incompatible, in which case we can stop right there, and not look at any others. It would therefore be handy to have a procedure which, with luck, might give us a decision before we have examined all pairs of fragments.

This is precisely how we arranged it. Define a graph K as follows. The nodes of K correspond to the fragments F_i and two nodes are adjacent if and only if the corresponding fragments are incompatible. Thus K is the graph whose adjacency matrix is $[a_{ij}]$. We want to assign a label, "0" (for "outside") or "I" (for "inside"), to each node of K, in such a way that adjacent nodes have dissimilar labels. In short, we seek a proper 2-colouring of the nodes of K. This will be possible if and only if K has no circuits of odd length (see [10] p. 170). The edges of K are presented to us one at a time as the pairs of fragments are examined for compatibility, and we want an algorithm that will stop as soon as a circuit of odd length is completed in the subgraph of K so far built up.

At any stage in the construction of K, the graph obtained so far will have a certain number of components. Thus, in addition to having a label "0" or "I", each node of K will be associated with a certain component, to which it belongs. As new edges are added, new components may form, or two components may join together to become one.

Whenever a new edge of K (i. e. a pair of incompatible fragments) is discovered, we look at its nodes. One of the following situations will apply and the stated action is taken:

(a) Neither node has appeared before: regard them as forming a new component, label one with '0' and one with 'I'.

(b) One node has been met before: label the other node in the opposite way, and put it in the same component.

(c) Both nodes have been met before and are in different components. There are two possibilities:

(i) They are labelled differently: make the two components into one. (This action is taken because the new edge joins the two components).

(ii) They are labelled similarly: make the two components into one, and reverse all the labels on the nodes of one of the previous components. (This is to make the allocation of labels correct across the new edge).

(d) Both nodes have been met before, and are in the same component. Again there are two possibilities.

(i) If the labels are different: breath a sigh of relief and do nothing.

(ii) If the labels are the same: an odd circuit has now been formed, and we need go no further. The graph G is non-planar.

The above "sub-algorithm" is quite easy to program, and besides giving the possibility of an early decision, it leads to a great saving in storage space. For each node of K we need only store the label (0 or 1) and the identification of the component to which it belongs.

This second algorithm for planarity may sound more complicated than the first; but it is better adapted for machine computation, and has been programmed by Fisher and Wing, by the author, and doubtless by many others.

7. Shortest paths and one-way streets.

We now turn to a completely different sort of problem. A common example of a network in real life is that of a system of highways, railroads, or other means of transportation between different points. For definiteness, let us think of a network of streets. The nodes of this network are the junctions where three or more streets meet, the edges are the streets between junctions, and associated with each one is a number-its length. There are many questions one can ask about such a network, one of the most obvious being "What is the shortest path between two given nodes?"

Let the nodes of the network be $A_1, A_2, \ldots, A_p$. If nodes A_i and A_j are joined by an edge, let d_{ij} denote the distance from A_i to A_j. For those pair of nodes which are not adjacent, i. e. not directly linked by a street, we put $d_{ij} = \infty$, where '∞' denotes some suitably large number. For greater generality we shall not assume that $d_{ij} = d_{ji}$, thus allowing the possibility of A_i and A_j being linked by two one-way streets of different length or, if one length is ∞, of being linked by a one-way street in one direction only. Naturally we take $d_{ii} = 0$ always. We now discuss the problem of finding the length of the shortest path between any two given nodes in terms of the matrix $D = [d_{ij}]$ just defined.

Several algorithms have been given for this, but the last word on the subject (at least as far as the matrix treatment is concerned) seems to have been said by T. C. Hu [7]. Consider the equation

$$d_{ij}^{(2)} = \min_k (d_{ik} + d_{kj}) . \tag{1}$$

Now $d_{ik} + d_{kj}$ is the length of the path from A_i to A_j passing through just one intermediate node A_k. Since we allow $k = j$, the direct distance d_{ij} is included in the quantities on the left of (1). Hence $d_{ij}^{(2)}$ is the shortest path between A_i and A_j using at most two edges. By comparison with the usual rule

$$c_{ij} = \sum_k a_{ik} b_{kj}$$

for matrix multiplication, we see that the matrix $[d_{ij}^{(2)}]$ is the "square" of the matrix D under a rather odd sort of multiplication. We can denote it by D^2 without creating overmuch confusion.

If we "square" D^2 in exactly the same way, we obtain a matrix whose (i, j)-element is the length of the shortest path from A_i to A_j using at most four edges. After N similar "squarings" in all, we obtain a matrix whose (i, j)-element is the length of the shortest path from A_i to A_j using at most 2^N edges. If $2^N > p - 1$, where p is the number of nodes in the network, this matrix gives us what we want; for no path can have more than $p - 1$ edges. (Unlike a walk, a path does not go through any node more than once).

This procedure in itself would be a reasonably convenient method for finding shortest paths, but a surprising and elegant result, described by Hu (due originally to Farbey, Land and Murchland [3]), is that we can do the whole thing with no more than two "squaring" operations only slightly different from that defined by (1). The trick is that, whenever we calculate an element $d_{ij}^{(2)}$ using (1), we immediately write it in place of the element d_{ij} in the matirx on which we are working, and subsequent calculations use these new values and not the original ones. Naturally we must specify the order in which we compute the elements $d_{ij}^{(2)}$, and in the first "squaring" we go from left to right along the top row, and then similarly with the other rows working downwards. When we "square" the matrix the second time we do the exact opposite; right to left along the bottom row, and then working up through the other rows. The matrix resulting from this two-stage process is the matrix of shortest distances that was obtained before.†

This algorithm is very convenient for calculation, both by hand and by digital computer. When we think of computers these days we usually have digital computers in mind, but we should not forget that there are such things as analogue computers, that are occasionally useful. It is possible to use analogue techniques on the shortest path problem for undirected networks (i. e. for which $d_{ij} = d_{ji}$). The "analogue computer" in question is not the usual electrical

---

†At the seminar it was pointed out to me by T. C. Hu that there now exists a one-stage process for finding the matrix of shortest distances. It is described in the following references:

S. Warshall, "A Theorem on Boolean Matrices", J. ACM, Vol. 9, pp. 11-12, 1962,

R. W. Floyd, "Algorithm 97, Shortest path", Comm. ACM, Vol. 5, p. 345, 1962,

and has appeared in Dr. Hu's recent book:

Integer Programming and Network Flows, Addison-Wesley, November 1969.

device, but a genuine home-grown, do-it-yourself model. All we need is some string, a ruler, and a pair of scissors. We cut lengths of string proportional to the lengths of the edges of the network, and knot their ends together so as to obtain a model of the network in which the nodes are represented (with pleasing etymological correctness) by the knots, and the edges by the strings between them. To find the shortest distance between any two nodes, we take hold of the corresponding knots and pull them apart as far as we can. The shortest path is given by those strings that are taut, and the distance between the knots is its length. (See [14]).

This "string algorithm" has, perhaps, rather more amusement value than practical utility, though it could be quicker than a hand calculation under favourable circumstances. It is, of course, quite unsuitable for digital computers, and it has the disadvantage that it will not work for directed networks. However, Klee [9] has shown that, by applying this string algorithm (or any other algorithm giving the same result) several times to a progressively modified undirected network, the shortest-path problem for a directed network can be solved. Since the only modifications to the network are the deletions of certain edges, the string algorithms is just the thing; all you have to do is to snip the appropriate pieces of string!

8. BFI algorithms

I cannot bring myself to conclude this paper without making some mention of BFI (Brute Force and Ignorance) algorithms. These are algorithms, devoid of any subtlety whatever, which simply keep thumping the problem on the back until it disgorges an answer. They are usually used only when there is no theoretical background to make possible a more sophisticated approach. A few examples will show what is meant.

If we have determined that a graph cannot be drawn on the plane, we may still ask "What is the simplest surface on which it <u>can</u> be drawn?" By 'simplest surface' we mean the surface of smallest genus. Let us first consider cubical graphs (in which all nodes have valency 3) and the problem of imbedding such a graph in an orientable surface. Such a graph, imbedded in a surface, will divide it into a number of regions. Let us walk round the boundary of each region in an anticlockwise direction (as seen from outside the surface) and put an arrow on each edge that we traverse, indicating the direction in which we traversed it. Then every edge will receive two arrows (clearly), and they will be in opposite directions. This means that if we look down on a node from above the surface, the arrows showing which way the three adjacent regions were traversed will look either as in figure 12(i) or figure 12(ii).

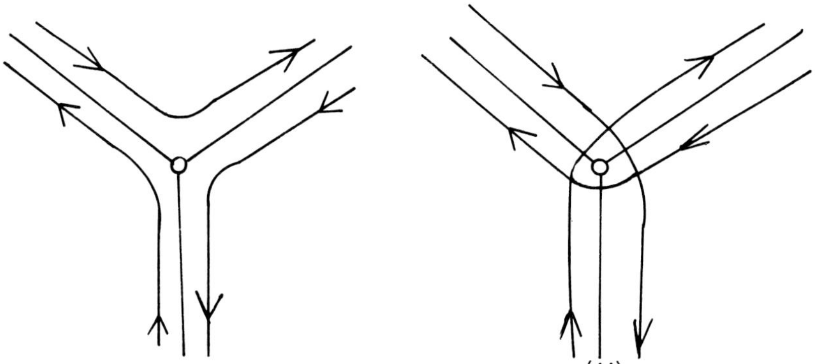

(i) Figure 12 (ii)

Conversely, if we construct a number of regions, bounded by edges, by walking round the graph in a number of circuits which cover each edge twice, once in each direction, then we can reconstruct the surface by pasting the regions together at their common edges. We make these circuits as follows. For each node we decide whether it is to be like figure (12(i) or like figure 12(ii), and label it 0 or 1 accordingly. Thus when we reach a given node by a certain edge, the label tells us by which edge to leave that node. Consequently, any allocation of 0's and 1's to the nodes gives us a set of circuits. We start along any edge, obey the traffic regulations at each node we come to, and carry on until we reach our starting point; and continue doing this until all edges have been traversed twice.

Let the number of nodes in the cubical graph be $q = 2k$ (it must be even). The number of edges is then $3k$. If the number of circuits constructed as above is f, then Euler's formula gives

$$f - 3k + 2k = 2 - 2p \quad \text{or} \quad p = (2 + k - f)/2$$

where p is the genus of the surface. The method is therefore to run through all the possible orientations of the nodes, find f for each one, and take note whenever we find a larger f than those found hitherto.

This kind of procedure is pure BFI; it just hammers away at all possibilities. By fixing the orientation of any one node we can cut the time of computation in half, but it is not easy to do much more than this.

Slightly more sophisticated than pure BFI algorithms are those that use "back-tracking", as described, for example, by Golomb and Baumert [5]. We can illustrate this by another method for imbedding a graph (any graph) in a surface (not necessarily orientable). Number the edges of the graph G with the integers $1, 2, 3, \ldots, q$ in any way. List all the circuits of G and express each as a q-dimensional vector of 0's and 1's, as described earlier. We now look for a set of vectors whose sum is $(2, 2, 2, \ldots, 2)$. These circuits will be the polygonal boundaries of pieces of surface, which can be stuck together to give the surface, since they include every edge twice.

If G has m circuits $\underline{x}_1, \underline{x}_2, \ldots, \underline{x}_m$ we start by finding the sum $\underline{x}_1 + \underline{x}_2$ (ordinary arithmetic now). We add $\underline{x}_3$, and if the resultant vector has no element '3', we add $\underline{x}_4$ to it, and so on. If the addition of $\underline{x}_i$ gives a vector with a '3' in it we have an inadmissible combination of circuits. We then backtrack to the previous sum, and try adding $\underline{x}_{i+1}$. If this fits, we carry on, if not, we try $\underline{x}_{i+2}$, and so on. If none of the subsequent circuits fit, we backtrack to the next previous sum, and so it goes on. In this way it is not necessary to try all 2^m combinations of "choose/don't choose this circuit". For example, if the first three circuits form an inadmissible combination, all the 2^{m-3} combinations which contain them will be rejected *en bloc.* Thus this type of program is a bit more subtle than the previous one, but not much.

BFI algorithms are often fairly easy to program for a computer; usually the crucial question is whether the computation can be effected in a reasonable length of time. I should like to conclude with a personal anecdote which illustrates this point. A few years ago I wrote a program for finding the number of Hamiltonian circuits in a given graph. This program used a back-tracking procedure, performing a walk around the edges of the graph, and backtracking whenever it came to a node previously encountered, unless this was the starting node and completed a circuit of length p. Every time a Hamiltonian circuit was found, a counter was incremented by 1.

This program worked very well on small graphs, and we decided to see if it would find the number of Hamiltonian circuits in Q_5, the 5-dimensional cube, which was (and still is) an open question. A slightly garbled version of what happened then has been given by Harary [6]. After running the program for a short while, a core dump was examined to see how far the program had progressed, and to estimate how long it would take to do the whole job. We estimated about 10 hours, and accordingly set the computer to run overnight unattended. During the night a tropical storm interrupted the power supply and the computer shut down. This was only a minor setback, and we would not, for that reason alone, have given up. The punch-line (which Harary omits) is that idle curiosity prompted us to look to see where the program had got to before being so abruptly terminated, and in doing so we discovered that we had made a rather serious error in calculating our previous estimate of the running time. Our revised estimate turned out to be more like 10 years! At this stage the project was abandoned!

REFERENCES

1. Dijkstra, E. W. A note on two problems in connexion with graphs, Numer. Math. 1(1959) 269-271.

2. Euler, L., Solutio Problematis ad Geometriam situs pertinentis. Commentarii Academiae Scientiarum Petropolitanae. vol VIII (1736) 128-140.

3. Farbey, B. A., Land, A. H., Murchland, J. D., The cascade algorithm for finding a minimum distance. Report LSE-TNT-19, London School of Economics, 1965.

4. Fisher, G. J., Wing, O., Computer recognition and extraction of planar graphs from the incidence matrix. IEEE Trans. Circuit Theory CT-13, 2(1966) 154-163.

5. Golomb, S. W., Baumert, L. D., Backtrack programming. J. Assoc. Comp. Mach. 12(1965) 516-524.

6. Harary, F., On the terminology of graph theory. Graph Theory and Theoretical Physics. Academic Press (1967).

7. Hu, T. C., Revised matrix algorithms for shortest paths. Siam J. Appl. Math. 15 (1967) 207-218.

8. Kirchhoff, G. Über die Auflösung der Gleichungen, auf welche man bei der Untersuchung der linearen Verteilung galvanische Ströme geführt wird. Annalen der Physik und Chemie, 72 (1947) 497-508.

9. Klee, V., A "string algorithm" for shortest path in directed networks. Operations Res. 12 (1964) 428-432.

10. König, D., Theorie der endlichen und unendlichen graphen. Leipzig 1936 (Reprinted, Chelsea 1950).

11. Kruskal, J. B., On the shortest spanning subtree of a graph and the travelling salesman problem. Proc. Amer. Math. Soc. 7 (1956) 24-50.

12. Kuratowski, C., Sur le probleme des courbes gauches en topologie. Fund. Math. 15 (1930) 271 - 283.

13. Mayeda, W., Seshu, S., Generation of trees without duplication. IEEE Trans. Circuit Theory CT-12 (1965) 181-185.

14. Minty, G. J., A comment on the shortest route problem. Operations Res. 5(1957) 724.

15. Read, R. C., Teaching graph theory to a computer. Recent Progress in Combinatorics. Academic Press. 1969.

16. Tutte, W. T., How to draw a graph. Proc. London Math. Soc. 13 (1963) 743 - 768.

17. Weinberg, L., Two new characterizations of planar graphs. Read at the Fifth Allerton Conference on Circuit and Systems Theory. University of Illinois.

18. Welch, J. T., A mechanical analysis of the cyclic structure of undirected linear graphs. J. Assoc. Comp. Mach. 13 (1966) 205-222.

On Eigenvalues and Colorings of Graphs

ALAN J. HOFFMAN

1. Introduction.

Let G be a graph (undirected, on a finite number of vertices, and no edge joining a vertex to itself). Let $A = A(G) = (a_{ij})$ be the adjacency matrix of G, i. e.,

$$a_{ij} = \begin{cases} 1 & \text{if } i \text{ and } j \text{ are adjacent vertices} \\ 0 & \text{if } i \text{ and } j \text{ are not adjacent vertices} \end{cases}.$$

For any real symmetric matrix A of order n, we will denote its eigenvalues arranged in descending order by

$$\lambda_1(A) \geq \ldots \geq \lambda_n(A) ,$$

and the eigenvalues in ascending order by

$$\lambda^1(A) \leq \ldots \leq \lambda^n(A).$$

(Thus $\lambda_i(A) = \lambda^{n-i+1}(A)$): If $A = A(G)$, we shall write $\lambda_i(G)$ and $\lambda^i(G)$ for $\lambda_i(A(G))$ and $\lambda^i(A(G))$ respectively.

Over the past ten years, there has been much work done on the question of relating geometric properties of G to the eigenvalues of $\lambda_i(G)$, and it had been my original intention to devote this talk to summarizing what has been accomplished since the survey given in the spring of 1967 [6]. Unfortunately, I could find no pedagogically sound way to organize this material. Instead, I will describe some observations I made during the summer connecting the eigenvalues of a graph with its coloring number and with some related concepts.

By this tactic, I hope that those who have never been previously exposed to any of the work relating eigenvalues to graphs will become convinced that there is some relevance.

We require some definitions. If G is a graph, its set of vertices is denoted by $V(G)$, its set of edges by $E(G)$. If $S \subset V(G)$, $S \neq \emptyset$, $G(S)$ is the graph such that $V(G(S)) = S$, $E(G(S))$ = the subset of $E(G)$ consisting of all edges both of whose vertices are in S. If G and H are graphs, $G \subset H$ if there is a subset $S \subset V(H)$ such that $G = H(S)$. If G is a graph on n vertices, and $|E(G)| = \binom{n}{2}$, then $V(G)$ is called a clique on n vertices and G is denoted by K_n. If G is a graph, $\bar{G}$ is the graph with $V(\bar{G}) = V(G)$, $E(\bar{G}) \cap E(G) = \emptyset$, $|E(\bar{G})| + |E(G)| = \binom{n}{2}$. The graph $\bar{G}$ is called the complementary graph of G, and satisfies $A(G) + A(\bar{G}) = J - I$, where J is the $n \times n$ matrix all of whose elements are unity. If $E(G) = \emptyset$, then $V(G)$ is called an independent set.

If i is a vertex of G, $d_i(G)$ is the number of vertices of G adjacent to i, and is called the valence of i. We also define

$$D(G) = \max_i d_i(G); \quad d(G) = \min_i d_i(G).$$

Note that $d_i(G) = \sum_j (A(G))_{ij}$.

A coloring of a graph G is a partitioning of $V(G)$ into independent sets. The coloring number of G is the smallest number of independent sets arising in such a partition and is denoted by $\gamma(G)$. More generally, let

$$V(G) = K^1 \cup \ldots . \cup K^r \cup I^1 \cup \ldots . \cup I^s \tag{1.1}$$

be a partition of $V(G)$ into subsets such that

(i) each K^i is a clique with at least two vertices, $i = 1, \ldots, r$;

(ii) each I^i is an independent set;

if such a decomposition (1.1) is possible, then G will be

said to admit an (r, s) decomposition.

The inspiration for the present investigation is an elegant result of Wilf [11]:

(1.2) $$\gamma(G) \leq 1 + \lambda_1(G) .$$

This upper bound is an improvement of part of a theorem of Brooks [3]

(1.3) $$\gamma(G) \leq 1 + D(G) .$$

In §2, (1.2) is proved and shown to imply (1.3).

To describe the results of §3 we need further definitions. Let $S_1 \cup \ldots \cup S_t$, $t \geq 2$ be a partition of $\{1, \ldots, n\}$ into non-empty subsets. For any real symmetric matrix A of order n, let $A_{ij}(i, j = 1, \ldots, t)$ be the submatrix of A with rows in S_i and columns in S_j. Aronzajn [1] has proved that if $t = 2$, $i_1 < |S_1|$, $i_2 < |S_2|$, then

(1.4) $$\lambda^1(A) + \lambda_{i_1+i_2+1}(A) \leq \lambda_{i_1+1}(A_{11}) + \lambda_{i_2+1}(A_{22}) .$$

With the help of a theorem of Wielandt [10] we prove that, for all $t \geq 2$, if $i_1 < |S_1|, \ldots, i_t < |S_t|$, then

(1.5) $$\lambda_{i_1+\ldots+i_t+1}(A) + \sum_{i=1}^{t-1} \lambda^i(A) \leq \sum_{k=1}^{t} \lambda_{i_k+1}(A_{kk}).$$

In §4, (1.5) is used to derive lower bounds on $\gamma(G)$ (more generally, on r and s in an (r, s) decomposition of G) in terms of the eigenvalues of G. For example,

(1.6) $$\frac{\lambda_1(G)}{|\lambda^1(G)|} + 1 \leq \gamma(G) .$$

We show in §5 that (1.6) is sharp in a number of interesting cases, and we also attempt to compare (1.6) with a lower bound for $\gamma(G)$ given in terms of $\{d_i(G)\}$ due to Bondy [2]. There does not seem to be an easy way to compare (1.6) with

[2], except in the case of regular graphs $(D(G) = d(G))$, where the comparison is favorable to (1.6).

In §6, we consider the cliquomatic number $\kappa(G)$ of a graph T, defined by $\kappa(G) = \gamma(\bar{G})$. Using a theorem of Lidskii [9], we derive from §4, lower bounds on $\kappa(G)$ in terms of the eigenvalues of G.

Finally, we mention in §7, results of a different kind: upper bounds for $\gamma(G)$ and $\kappa(G)$ as functions of (respectively) the number of eigenvalues of G each of which is at most -1, and the number of nonnegative eigenvalues of G. Also, we state that for each k, there exists upper bounds for r and s in an (r, s) decomposition of G, where the respective upper bounds depend on $\max(\lambda_k(G), -\lambda^1(G))$.

2. Wilf's Theorem

For the sake of amusement, we will give a proof slightly different from Wilf's, using the maximum characterization of eigenvalues of a real symmetric matrix in lieu of his appeal to the Perron-Frobenius theory.

Let A be a real symmetric matrix of order n, B a principal submatrix of W. Then from the maximum principle we infer

$$\lambda_1(A) \geq \lambda_1(B), \tag{2.1}$$

$$\lambda^1(B) \geq \lambda^1(A) \tag{2.2}$$

(we will use (2.2) in later sections). Further

$$\lambda_1(A) \geq \frac{1}{n} \sum_{i,j} a_{ij} . \tag{2.3}$$

For, let $u = (1, \ldots, 1)$. Then the right-hand side of (2.3) is the Rayleigh quotient $(Au, u)/(u, u)$, and every Rayleigh quotient formed from A is at most $\lambda_1(A)$. This argument is contained in [4].

If $A = A(G)$, we infer from (2.3) that

(2.4) $$\lambda_1(G) \geq \frac{1}{n} \sum_i d_i(G) \geq d(G) .$$

Further

(2.5) $$\lambda_1(G) \leq D(G) ,$$

since by Gersgorin's theorem, $\lambda_1(G) \leq \max_i \sum_i a_{ij} = D(G)$.

Comment: This also follows from the Perron-Frobenius Theory- viz $\max \lambda_1 \leq \min(\max_j \Sigma a_{ij}, \max_i \Sigma a_{ij})$ which reduced to the same.

To prove (1.2) we first observe that there must exist a subgraph $H \subset G$ such that $d(H) \geq \gamma(G)-1$ (otherwise we would contradict that $\gamma(G)$ is the coloring number of G). From (2.4), we have

(2.6) $$\lambda_1(H) \geq d(H) \geq \gamma(G)-1.$$

From (2.1), $\lambda_1(H) \leq \lambda_1(G)$. Combining this with (2.6), we infer (1.2). Next, inequality (1.3) follows from (1.2) and (2.5)

3. A Lemma on Partitioned Matrices.

Let A be a real symmetric matrix of order n, and let $S_1 \cup \ldots \cup S_t$, $t \geq 2$, be a partition of $\{1, \ldots, n\}$ into nonempty subsets. If $i_k < |S_k|$, $k = 1, \ldots, t$, then

(3.1) $$\lambda_{1+\sum_{k=1}^{t} i_k}(A) + \sum_{i=1}^{t-1} \lambda^i(A) \leq \sum_{k=1}^{t} \lambda_{i_k+1}(A_{kk}).$$

Proof: We shall prove the lemma by induction on t. In case $t = 2$, the lemma is Aronzajn's inequality [1] (see (1.4)). Assume therefore that the lemma has been proved for $t-1$. Let $T = S_1 \cup \ldots \cup S_{t-1}$ and let $A[T]$ be the principal submatrix of A formed by rows and columns in T. By the induction hypothesis

(3.2) $$\lambda_{1+\sum_{k=1}^{t-1} i_k}(A[T]) + \sum_{i=1}^{t-2} \lambda^i(A[T]) \leq \sum_{k=1}^{t-1} \lambda_{i_k+1}(A_{kk}) .$$

Let $\tilde{x}^1, \tilde{x}^2, \ldots, \tilde{x}^{|T|}$ be an orthonormal set of eigenvectors of $A[T]$, so that

$$A[T]\tilde{x}^j = \lambda^j(A[T])\tilde{x}^j, \quad j = 1, \ldots, |T|. \tag{3.3}$$

Let $x^1, \ldots, x^{|T|}$ be the vectors with n coordinates obtained respectively from $\tilde{x}^1, \ldots, \tilde{x}^{|T|}$ by putting 0 for all coordinates x_ℓ^j, $\ell \in T$. Let B be the matrix representing the same linear operator as A, with respect to the orthonormal basis $x^1, \ldots, x^{|T|}$ and unit vectors v^ℓ, $\ell \in T$. Then B has the same eigenvalues as A, and $B[T]$ is diagonal.

Let U be the set of indices $j \in T$ such that $\{B[T]_{jj}\}_{j \in U}$ consists of $\{\lambda^{t-1}, \ldots, \lambda^{|T|}\}$. Let $N = U \cup S_t$. Thus for any $j_1 < |U|$ and $j_2 < |S_t|$, by Aronzajn's inequality (1.4)

$$\lambda^1(B[N]) + \lambda_{j_1+j_2+1}(B[N]) \le \lambda_{j_1+1}(B[U]) + \lambda_{j_2+1}(A_{tt}).$$

Since $|U| = |T| - t+2$ and $i_k \le |S_k| - 1$, $k = 1, 2, \ldots, t-1$, $\sum_{k=1}^{t-1} i_k \le \sum_{k=1}^{t-1} |S_k| - t+1 = |T| - t+1 < |U|$, setting $j_1 = \sum_{k=1}^{t-1} i_k$ and $j_2 = i_t$, we obtain

$$\lambda_{1+\sum_{k=1}^{t} i_k}(B[N]) + \lambda^1(B[N]) \le \lambda_{1+\sum_{k=1}^{t-1} i_k}(B[U]) + \lambda_{i_t+1}(A_{tt}). \tag{3.4}$$

Note that by our construction of U,

$$\lambda_{1+\sum_{k=1}^{t-1} i_k}(B[U]) = \lambda_{1+\sum_{k=1}^{t-1} i_k}(A[T]),$$

and

$$\lambda^i(A[T]) = \lambda^i(B[T]) \quad i = 1, \ldots, |T|.$$

Hence, adding (3.2) and (3.4), we obtain

$$\lambda_{1+\sum_{k=1}^{t} i_k}(B[N]) + \lambda^1(B[N]) + \sum_{i=1}^{t-2} \lambda^i(B[T] \le \sum_{k=1}^{t} \lambda_{i_k+1}(A_{kk}). \tag{3.5}$$

We now invoke a lemma due to Wielandt [10]: Let C be a real symmetric matrix of order n, and let $1 \le j_1 < j_2 < \ldots < j_r \le n$ then

$$\min_{S_{j_1} \subset \ldots \subset S_{j_r}} \max_{\substack{x_\ell \in S_{j_\ell} \\ (x_\ell, x_m) = \delta_{\ell m}}} \sum_{\ell=1}^{r} (Cx_\ell, x_\ell) = \sum_{\ell=1}^{r} \lambda^{j_\ell}(C) \tag{3.6}$$

(In the left side of (3.6), S_ℓ stands for a linear subspace of dimensional ℓ). Let $\{\tilde{y}^i\}_{i \in N}$ be an orthonormal set of eigenvectors of $B[N]$ so that

$$B[N]\, \tilde{y}^i = \lambda^i(B[N]), \quad i = 1, \ldots, |N|, \tag{3.7}$$

and let y^i be the vector with n coordinates obtained from $\tilde{y}^i$ by putting 0 for all coordinates y^i_ℓ, $\ell \notin N$. Then $\{x^1, \ldots, x^{t-2}, y^1, \ldots, y^{n-t+2}\}$ are an orthonormal set of vectors. For $i = 1, \ldots, t-2$, let T_i be the vector space spanned by $\{x^1, \ldots, x^i\}$. For $i = t-1, \ldots, n$, let T_i be the vector space spanned by $\{x^1, \ldots, x^{t-2}, y^1, y^2, \ldots, y^{i-t+2}\}$. Note $\dim T_i = i$ in all cases.

Let $V = \{1, \ldots, t-2, t-1, n - \sum_{k=1}^{t} i_k\}$. Then

$$\max_{\substack{x_\ell \in T_\ell \\ (x_\ell, x_m) = \delta_{\ell m}}} \sum_{\ell \in V} (Bx_\ell, x_\ell) = \sum_{i=1}^{t-2} \lambda^i(B[T]) + \lambda^1(B[N]) + \lambda_{1+\sum_{t=1}^{t} i_k}(B[N]), \tag{3.8}$$

from (3.3), (3.7) and the construction of the $\{T_i\}_{i \in V}$. By (3.6),

$$\lambda_{1+\sum_{k=1}^{t} i_k}(B) + \sum_{i=1}^{t-1} \lambda^i(B) \le \max_{\substack{x_\ell \in T_\ell \\ \{x_\ell\}_{\ell \in V} \\ \text{orthonormal}}} \sum_{\ell \in V} (Bx_\ell, x_\ell). \tag{3.9}$$

Combining (3.8), (3.9) and (3.5) yields (3.1).

4. Lower Bounds for Coloring and (r, s) Decompositions.

It is now a simple matter to apply (3.1) and derive lower bounds for $\gamma(G)$ and (r, s) decompositions. We first prove: if $\gamma = \gamma(G) \geq 2$, then

$$\lambda_1(G) + \sum_{i=1}^{\gamma-1} \lambda^i(G) \leq 0 . \tag{4.1}$$

Proof: By hypothesis, $V(G)$ can be partitioned into non-empty subsets $S_1 \cup \ldots \cup S_\gamma$ such that S_k is an independent set, $k = 1, 2, \ldots, \gamma$. Consequently, $\lambda_1(A(G)_{kk}) = 0$, $k = 1, \ldots, \gamma$. Apply (3.1) with each $i_k = 0$, and (4.1) follows.

Before proceeding further, note that for $t \geq 2$, $\lambda_1(K_t) = t - 1$, $\lambda^1(K_t) = \ldots = \lambda^{t-1}(K_t) = -1$. In particular, it follows from (2.2) that if G has at least one edge (so $\gamma(G) \geq 2$, and $K_2 \subset G$), $\lambda^1(G) \leq -1$. We now infer from (4.1)

$$\gamma(G) \geq \frac{\lambda(G)}{-\lambda^1(G)} + 1, \quad \text{if } \gamma(G) \geq 2 . \tag{4.2}$$

Proof: By (4.1), we have

$$\lambda_1(G) + (\gamma-1)\lambda^1(G) \leq 0 .$$

Using the fact that $\lambda^1(G) < 0$, (4.2) follows. We next prove: If G has an (r, s) decomposition, then

$$\lambda_{r+1}(G) + \sum_{i=1}^{r+s-1} \lambda^i(G) \leq -r . \tag{4.3}$$

Proof: Recalling (1.1), we see that, if we use $i_k = 1$ for the indices k referring to cliques, and $i_\ell = 0$ for the indices ℓ ℓ referring to independent sets, and use $\lambda_2(K_t) = -1$ if $t \geq 2$, then (3.1) becomes (4.3).

In particular, we infer from (4.3) that

$$\lambda_{r+1}(G) + (r+s-1)\lambda^1(G) \leq -r , \qquad \text{or}$$

$$\frac{\lambda_{r+1}(G) + r}{-\lambda^1(G)} + 1 - r \leq s . \tag{4.4}$$

5. Further Comments on the Lower Bound for $\gamma(G)$.

The lower bound for $\gamma(G)$ given by (4.2) is sharp in a number of interesting cases. For example, it is known [5] that $\gamma(G) = 2$ if and only if $\lambda^1(G) + \lambda_1(G) = 0$. In this case, the lower bound given by (4.2) is exact. If G is an odd polygon, then $\gamma(G) = 3$, $\lambda_1(G) = 2$, $-\lambda^1(G) = 2 - \epsilon$, where $0 < \epsilon \leq 1$. Thus the right hand side of (4.2) becomes $1 + \alpha$, when $1 < \alpha \leq 2$. But $\gamma(G) \geq 1+\alpha$ implies $\gamma(G) \geq -[-(1+\alpha)]$, where $[x]$ is the largest integer not exceeding x. Thus, in this case, (4.2) is also sharp. If G has n independent sets, each consisting of m vertices, such that any pair of vertices of different independent sets are adjacent, then $\gamma(G) = n$, $\lambda_1(G) = m(n-1)$, $\lambda^1(G) = -m$, and again (4.2) is sharp.

If M is a (0, 1) matrix such that every row sum is k, every column sum is k, $k \geq 2$ let $G = G(M)$ be the graph whose vertices are the 1's in M, with two vertices adjacent if the corresponding 1's are in the same row or column. Then it has been shown [6] that $\lambda_1(G) = 2k-2$, $\lambda^1(G) = -2$, so (4.2) shows that $\gamma(G) \geq k$. On the other hand, by a theorem of Konig [8] there exist permutation matrices $P_1, \ldots, P_k$ such that $M = P_1 + \ldots + P_k$. The 1's occurring in each P_i form an independent set, so $\gamma(G) \leq k$. Thus, (4.2) is sharp in this case.

It is also interesting to observe the implication of (4.2) when one knows an upper bound for $\gamma(G)$. For instance if G is planar then $\gamma(G) \leq 5$. By (4.2) this implies

$$-\lambda^1(G) \geq \frac{1}{4}\lambda_1(G) \quad \text{if } G \text{ is planar}.$$

Further $-\lambda^1(G) \geq \frac{1}{3}\lambda_1(G)$ if G is planar and the four color hypothesis is true. Similar remarks can be made, of course, for graphs imbedded on surfaces of higher genus.

Bondy [2] has given a lower bound for $\gamma(G)$ in terms of the $\{d_i(G)\}$. It is difficult to compare his lower bound with (4.2) except in the case where G is regular (of valence d). Then

$$\gamma(G) \geq \frac{n}{n-d}. \tag{5.1}$$

We will show (4.2) is a better bound, by proving

(5.2) $$\frac{\lambda_1(G)}{-\lambda^1(G)} + 1 \geq \frac{n}{n-d} \quad .$$

To prove (4.2) we recall from §2 that $\lambda_1(G) = d$, so (5.2) holds if and only if

(5.3) $$0 \leq (n-d) + \lambda^1(G) \; .$$

Write $J = (J - A) + A$, where $A = A(G)$, and observe that since G is regular (and therefore A commutes with J), the eigenvalues of J are the sums of corresponding eigenvalues of J $J-A$ and A. Clearly (see §1), $\lambda_1(J) = n$, $\lambda_1(J-A) = n-d$, $\lambda_1(A) = d$. Also,

$$0 = \lambda_1(J) = \lambda_i(J-A) + \lambda^{i-1}(A) \qquad i = 2, \ldots, n \, .$$

In particular, setting $i = 2$, we have

(5.4) $$0 = \lambda_2(J-A) + \lambda^1(A) \; .$$

But $\lambda_2(J-A) \leq \lambda_1(J-A) = n-d$. Combined with (5.4), this yields (5.3). In fact, if $\bar{G}$ is connected, then $\lambda_2(J-A) < \lambda_1(J-A)$, so (4.2) is always at least as good a bound as (5.1) for regular graphs, and is a better bound if the complementary graph is connected.

6. Lower Bounds on $\kappa(G)$.

For any graph G, define $\kappa(G)$ to be the smallest number of cliques whose union contains all vertices of G. Then $\kappa(G) = \gamma(\bar{G})$. We shall prove that, if $\kappa = \kappa(G) \geq 2$, and $|V(G)| = n$ then

(6.1) $$n - \kappa \leq \sum_{i=1}^{\kappa} \lambda_i(G) \; .$$

Note that if $\kappa(G) \geq 2$, G contains two non-adjacent vertices- thus $A(G)$ contains a 2×2 principal submatrix which is 0. By the interlacing theorem, $\lambda_2(G) \geq 0$, hence $1 + \lambda_2(G) > 0$. Since the right-hand side of (6.1) is bounded from above by $\lambda_1(G) + (\kappa - 1)\lambda_2(G)$, we infer from (6.1) that

(6.2) $$\kappa(G) \geq \frac{n+\lambda_2(G) - \lambda_1(G)}{1+\lambda_2(G)} .$$

To prove (6.1), we recall a theorem of Lidskii [9]: if A, B and C are real symmetric matrices of order n, A = B + C. $1 \leq i_1 < i_2 < \ldots < i_r \leq n$, then

(6.3) $$\sum_{\ell=1}^{r} \lambda_{i_\ell}(A) \leq \sum_{\ell=1}^{r} \lambda_{i_\ell}(B) + \sum_{\ell=1}^{r} \lambda_{i}(C) .$$

Let $\kappa = \kappa(G) = \gamma(\bar{G})$, and write $J-I = A(G) + A(\bar{G})$. Let $r = k$, $t_1 = 1$, $t_k = n-1, \ldots, t_2 = n-k+2$. From (6.3), we infer

(6.4) $$(n-1) - (\kappa-1) \leq \lambda_1(\bar{G}) + \sum_{i=1}^{k-1} \lambda^{i}(\bar{G}) + \sum_{i=1}^{\kappa} \lambda_i(G) .$$

But from (4.1)

(6.5) $$\lambda_1(\bar{G}) + \sum_{i=1}^{\kappa-1} \lambda^{i}(\bar{G}) \leq 0 .$$

Combining (6.4) and (6.5), we infer (6.1).

7. Further Upper Bounds on $\gamma(G)$, $\kappa(G)$ and (r, s) Decompositions of G.

Let $M(G)$ = the number of eigenvalues of G, each of which is at most -1, and let $m(G)$ = the number of non-negative eigenvalues of G. Then one can prove there exist functions f and g such that

(7.1) $$\gamma(G) \leq f(M(G)) ,$$

(7.2) $$\kappa(G) \leq g(m(G)).$$

We conjecture that $f(G) = 1 + M(G)$, but have not succeeded in proving this.

Let $k > 1$. Using the results of [7], we can prove there exists functions f_k and g_k such that G has an (r, s) decomposition, where

(7.3) $$r \leq f_k(\max(\lambda_k(G), -\lambda^1(G)) ,$$

(7.4) $$s \leq g_k (\max (\lambda_k(G), -\lambda^1(G))).$$

We conjecture that f_k can be made a function of $\lambda^1(G)$ alone, g_k a function of λ^k alone, but have not yet succeeded in proving this.

We are grateful to Donald Newman, Robert Thompson and Herbert Wilf for useful conversations about this material.

REFERENCES

1. Aronzajn, N. "Rayleigh-Ritz and A. Weinstein Methods for Approximation of Eigenvalues. I. Operators in a Hilbert Space," Proc. Nat. Acad, Sci., U. S. A., 34(1948), 474-480.

2. Bondy, J. A. "Bounds for the Chromatic Number of a Graph," J. Combinatorial Theory, 7(1969), 96-98.

3. Brooks, R. L.. "On Colouring the Nodes of a Network, Proc. Cambridge Philos. Soc., 3791941), 194-197.

4. Collatz, L. and Sinogowitz, U., "Spektren endlicher Graphen," Abh. Math. Sem. Univ. Hamburg, 21(1957), 63-77.

5. Hoffman, A. J., "On the Polynomial of a Graph," Amer. Math. Monthly, 70 (1963), 30-36.

6. Hoffman, A. J., "The Eigenvalues of the Adjacency Matrix of a Graph," in Combinatorial Mathematics and its Applications, edited by R. C. Bose and T. C. Dowling, University of North Carolina Press, Chapel Hill, 1969, 578-584.

7. Hoffman, A. J., "$-1-\sqrt{2}$?" to appear in Proceedings of the Calgary International Conference on Combinatorial Structures and their Applications, Canada in June 1969.

8. Kőnig, D., "Theorie der Endlichen und Unendlichen Graphen," Chelsea, New York, 1950, 170-178.

9. Lidskii, V. B., "The Proper Values of the Sum and Product of Symmetric Matrices," Doklady Akad. Nauk. SSSR (N. S.) 75(1950, 769-772. Translation by C. D. Benster, National Bureau of Standards (Washington) Report 2248, 1953.

10. Wielandt, H., "An Extremum Property of Sums of Eigenvalues," Pacific J. Math. 5(1955), 633-638.

11. Wilf, H. S., "The Eigenvalues of a Graph and its Chromatic Number," J. London Math. Soc. 42(1967), 330-332.

Work supported in part by the office of Naval Research under Contract No. Nonr-3775(00).

Blocking Polyhedra

D. R. FULKERSON

1. Introduction

It is well known that the permutation matrices can be characterized geometrically as the extreme points of the convex polyhedron of all doubly stochastic matrices. A new result of this paper (Sec. 6) is that the permutation matrices can also be characterized geometrically as the extreme points of the following (unbounded) convex polyhedron. Let ξ_{ij} be a variable associated with cell i, j of an n by n array, and consider the linear inequalities:

$$\sum_{\substack{i \in I \\ j \in J}} \xi_{ij} \geq |I| + |J| - n, \quad \text{all } I, J \subseteq \{1, \ldots, n\}, \tag{1.1}$$

$$\xi_{ij} \geq 0, \quad \text{all } i, j \in \{1, \ldots, n\}. \tag{1.2}$$

Each extreme point of the polyhedron $\mathcal{P}$ defined in R^{n^2} by (1.1) and (1.2) is a permutation matrix $x = (\xi_{ij})$. Moreover, almost all of the inequalities (1.1) that correspond to positive right hand sides are essential in defining $\mathcal{P}$. Thus we have another, albeit more complicated, geometric representation of the permutation matrices.

The results of this paper were arrived at partly in an attempt to understand better the inequalities (1.1). The basic underlying geometric fact (Theorem 2.1) appears to be a certain polarity between members of the class of all (unbounded) convex polyhedra defined by linear inequalities of the form

$$(1.3) \qquad Ax \geq 1 ,$$

$$(1.4) \qquad x \geq 0 ,$$

where A is a nonnegative matrix, $0 = (0, \ldots, 0)$, and $1 = (1, \ldots, 1)$. We call a polar pair of this class a blocking pair of polyhedra, because the geometric theory developed here has intimate connections with the notion of a blocking pair of "clutters" defined on a finite set E [6, 9, 10]. (A clutter on E is a family of noncomparable subsets of E.) From this point of view, the present paper may be regarded as a continuation of [6, 9, 10]. In particular, it is found that the "length-width" inequality and the "max-flow min-cut" equality, studied in [10] for a blocking pair of clutters, are always valid for a blocking pair of polyhedra (Theorem 3.1).

I should like to thank Jon Folkman for helpful discussion concerning the proof of Theorem 2.1.

2. A Polarity.

Let A be an m by n nonnegative matrix, and consider the convex polyhedron

$$(2.1) \qquad \mathcal{B} = \{b \in R^n_+ \mid Ab \geq 1\}.$$

Here 1 denotes the m-vector all of whose components are 1 and R^n_+ is the nonnegative orthant of R^n. The polyhedron $\mathcal{B}$ is the vector sum of the convex hull of its extreme points and the nonnegative orthant:

$$(2.2) \qquad \mathcal{B} = \text{conv. hull} (\{b^1, b^2, \ldots, b^r\}) + R^n_+ ,$$

where $b^1, \ldots, b^r$ are the extreme points of $\mathcal{B}$.

We say that a row vector a^i of the matrix A is *inessential* if it dominates a convex combination of other rows of A, i.e., the inequality $a^i \geq \Sigma_{j=1}^m \alpha_j a^j$ holds for some $\alpha_1 \geq 0, \ldots, \alpha_m \geq 0$ satisfying $\alpha_i = 0$, $\Sigma_{j=1}^m \alpha_j = 1$; otherwise the row a^i is *essential*. It is a consequence of the

Farkas lemma that an inequality of (2.1) may be dropped in the definition of $\mathcal{B}$ if and only if the corresponding row of A is inessential. Accordingly we may suppose without loss of generality that all rows of A are essential. We call such an A <u>proper</u>, and include in this definition the degenerate cases (i) A is a one-rowed zero matrix ($\mathcal{B}$ is empty), and (ii) A has no rows ($\mathcal{B} = R^n_+$).

Let

$$\hat{\mathcal{B}} = \{a \in R^n_+ \mid a \cdot \mathcal{B} \geq 1\} . \tag{2.3}$$

We call $\hat{\mathcal{B}}$ the <u>blocker</u> of $\mathcal{B}$. Note that if $\mathcal{B}$ is empty, then $\hat{\mathcal{B}} = R^n_+$, and if $\mathcal{B} = R^+_n$, then $\hat{\mathcal{B}}$ is empty. Theorem 2.1 below shows that the blocking relation is a polarity on the dlass of all convex polyhedra of the form (2.1).

<u>Theorem 2.1.</u> Let the m by n matrix A be proper with rows $a^1, \ldots, a^m \in R^n_+$. Let $\mathcal{B} = \{b \in R^n_+ \mid Ab \geq 1\}$ have extreme points $b^1, \ldots, b^r$, let B be the matrix having rows $b^1, \ldots, b^r$, and let $\mathcal{A} = \{a \in R^n_+ \mid Ba \geq 1\}$. Then (i) $\hat{\mathcal{B}} = \mathcal{A}$; (ii) B is proper and the extreme points of $\mathcal{A}$ are $a^1, \ldots, a^m$; (iii) $\hat{\mathcal{A}} = \mathcal{B}$.

Theorem 2.1 can be deduced from standard results about polar cones. We give a direct proof below.

<u>Proof.</u> We first prove (i). Suppose $a \in \mathcal{A}$. Thus $b^1 \cdot a \geq 1, \ldots, b^r \cdot a \geq 1$. If $b \in \mathcal{B}$, then by (2.2) we have $b = \Sigma_{i=1}^r \alpha_i b^i + z$, where $z \geq 0$, $\alpha_i \geq 0$, $\Sigma_{i=1}^r \alpha_i = 1$. Thus $a \cdot b \geq \Sigma_{i=1}^r \alpha_i (a \cdot b^i) \geq 1$. Hence $a \in \hat{\mathcal{B}}$, and $\mathcal{A} \subseteq \hat{\mathcal{B}}$. Conversely, if $a \in \hat{\mathcal{B}}$, then $a \cdot b \geq 1$ for all $b \in \mathcal{B}$, and in particular, $a \cdot b^1 \geq 1, \ldots, a \cdot b^r \geq 1$. Thus $a \in \mathcal{A}$, and $\hat{\mathcal{B}} \subseteq \mathcal{A}$. Hence $\mathcal{A} = \hat{\mathcal{B}}$.

To show that B is proper, suppose $b^1 \geq \Sigma_{i=2}^r \alpha_i b^i$, where $\alpha_i \geq 0$, $\Sigma_{i=2}^r \alpha_i = 1$. Let $y = \Sigma_{i=2}^r \alpha_i b^i$. Then $b^1 = y + z$, $z \geq 0$. If $z = 0$, then clearly b^1 is not extreme. If $z \neq 0$, then $y + 1/2z$, $y + 3/2z$ are distinct points of $\mathcal{B}$ whose average is b^1, again contradicting the fact that b^1 is an extreme point of $\mathcal{B}$. Hence B is proper.

Let $C = \text{conv. hull}(\{a^1, \ldots, a^m\})$. We shall show that $C + R^n_+ = \mathcal{A}$. To this end, suppose first that $x \in C + R^n_+$, so that $x = \Sigma_{i=1}^{m} \alpha_i a^i + z$, $z \geq 0$, $\alpha_i \geq 0$, $\Sigma_{i=1}^{m} \alpha_i = 1$. Then $b^j \cdot x = b^j \cdot \Sigma_{i=1}^{m} \alpha_i a^i + b^j \cdot z \geq 1$. Thus $x \in \mathcal{Q}$, and $\mathcal{C} + R_+^n \subseteq \mathcal{Q}$. If equality does not hold here, then by the separating hyperplane theorem, there is a $b \in R^n$ and an $\alpha \in R$ such that $b \cdot x \geq \alpha$ for all $x \in \mathcal{C} + R_+^n$, whereas $b \cdot a < \alpha$ for some $a \in \mathcal{Q}$. Since $b \cdot x \geq \alpha$ for all $x \in \mathcal{C} + R_+^n$, we must have $b \geq 0$. Thus $\alpha > b \cdot a \geq 0$. Hence $(1/\alpha)b \cdot x \geq 1$ for all $x \in \mathcal{C} + R_+^n$, and in particular, $(1/\alpha)b \cdot a^i \geq 1$, $i = 1, 2, \ldots, m$. Thus $(1/\alpha)b \in \mathcal{B}$ and hence $(1/\alpha)b = \Sigma_{i=1}^{r} \beta_i b^i + z$, where $z \geq 0$, $\beta_i \geq 0$, $\Sigma_{i=1}^{r} \beta_i = 1$. But then

$$1 > \frac{1}{\alpha} b \cdot a = \sum_{i=1}^{r} \beta_i b^i \cdot a + z \cdot a \geq 1,$$

a contradiction. Hence $\mathcal{C} + R_+^n = \mathcal{Q}$. It then follows that the row vectors $a^1, \ldots, a^m$ of A are the extreme points of $\mathcal{C}$. For suppose a^1, say, is not extreme. Then $a^1 = (1/2)(x + y)$, where $x = \Sigma_{i=1}^{m} \alpha_i a^i + u \neq a^1$, $y = \Sigma_{i=1}^{m} \beta_i a^i + v \neq a^1$, $\alpha_i \geq 0$, $\beta_i \geq 0$, $\Sigma_{i=1}^{m} \alpha_i = 1$, $\Sigma_{i=1}^{m} \beta_i = 1$, and $u \geq 0$, $v \geq 0$. Moreover, $\alpha_1 + \beta_1 < 2$, since $x \neq a^1$, $y \neq a^1$. We have

$$a^1 = \sum_{i=1}^{m} \frac{(\alpha_i + \beta_i)}{2} a^i + (1/2)(u + v).$$

Let $(1/2)(\alpha_i + \beta_i) = \gamma_i$. Then $\gamma_i < 1$ and

$$a^1 \geq \sum_{i=2}^{m} \frac{\gamma_i}{1-\gamma_1} a^i, \quad \frac{\gamma_i}{1-\gamma_1} \geq 0, \quad \sum_{i=2}^{m} \frac{\gamma_i}{1-\gamma_1} = 1.$$

This contradicts the assumption that A is proper, and finishes the proof of (ii).

Part (iii) of Theorem 2.1 now follows from (i) and (ii).

We call the matrix B of Theorem 2.1 the <u>blocking</u> matrix of A. The blocking matrix of B is then A.

An example illustrating Theorem 2.1 in R^2 is shown in Fig. 1 below.

It follows from Theorem 2.1 that if we are given the matrix A that defines $\mathcal{B}$, then the blocking matrix B

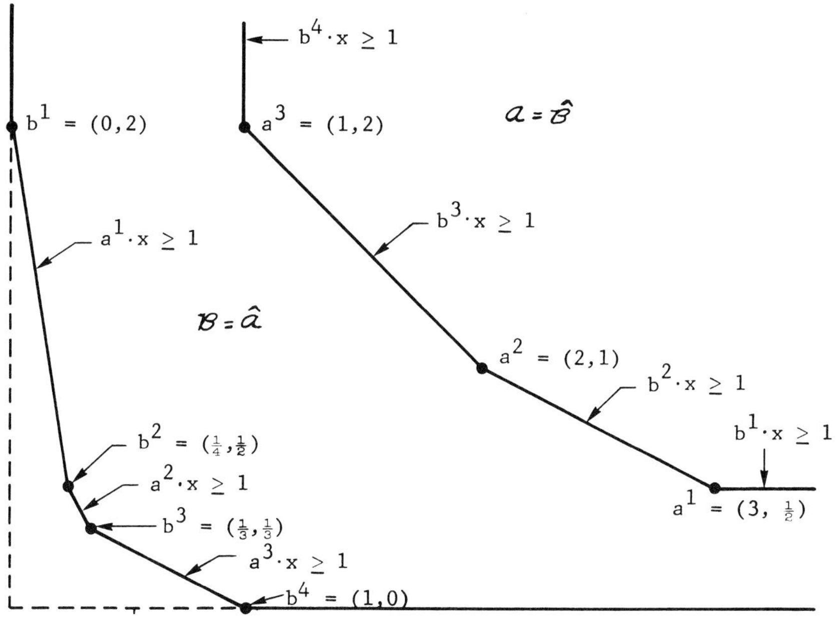

$$A = \begin{bmatrix} 3 & \frac{1}{2} \\ 2 & 1 \\ 1 & 2 \end{bmatrix} \qquad B = \begin{bmatrix} 0 & 2 \\ \frac{1}{4} & \frac{1}{2} \\ \frac{1}{3} & \frac{1}{3} \\ 1 & 0 \end{bmatrix}$$

Fig. 1

defining $\mathcal{G}$ can be determined by the following straightforward but exceedingly tedious process. Append the n by n identity matrix to A, and then find an n by n nonsingular submatrix $\bar{A}$ of the matrix thus obtained. Next solve the linear system of equations having $\bar{A}$ as coefficient matrix and having right hand side 1 or 0 according as the corresponding row of $\bar{A}$ belongs to A or to I. If the resulting solution b satisfies $b \geq 0$, $Ab \geq 1$, then b is a row of B. All rows of B are obtainable in this way.

The case in which A is a (0, 1)-matrix is of particular interest for extremal combinatorics. The assumption that A is proper is then equivalent to saying that A is the m by n incidence matrix of a clutter of m subsets of a set of n elements (no row of A contains another row of A). Thus Theorem 2.1 provides a way of characterizing the subsets comprising an arbitrary clutter as the extreme points of a convex polyhedron. If A is the incidence matrix of a clutter, then the incidence matrix b(A) of the blocking clutter has as rows all (0, 1)-vectors with n components that have inner product at least 1 with all rows of A and that are minimal with respect to this property [10]. It is not hard to see that each row of the matrix b(A) will then be a row of matrix B in Theorem 2.1. In general, B will have many other rows. But there are significant classes of clutters for which b(A) = B. For example, if A is the incidence matrix of all simple paths joining two distinguished nodes of a graph having n edges, then b(A) is equal to B, and hence B is the incidence matrix of all cuts separating the two nodes. (By the duality asserted in Theorem 2.1, we could of course start with b(A) and obtain A.)

In the context of Theorem 2.1, Lehman's interesting paper [10] can be viewed as a study of clutters A for which b(A) = B. Generalizing from the example above, he shows that this situation holds for a clutter A if and only if the max-flow min-cut equality holds for A and b(A), or if and only if the length-width inequality holds for A and b(A). Here, in analogy with the example of paths and cuts in a graph, the max-flow min-cut equality is said to hold for a (0, 1)-matrix A correponding to a clutter, and the matrix

$b(A)$ having rows $b^1, \ldots, b^r$, if and only if for every $w \in R^n_+$, it is true that in the linear program

$$(2.4) \qquad yA \leq w, \quad y \geq 0, \quad \max 1 \cdot y,$$

we have

$$(2.5) \qquad \max 1 \cdot y = \min_{1 \leq j \leq r} b^j \cdot w.$$

Similarly, the length-width inequality is said to hold for a (0, 1)-matrix A, whose rows $a^1, \ldots, a^m$ correspond to the sets of a clutter, and the matrix $b(A)$, if and only if for every $\ell \in R^n_+$, $w \in R^n_+$, we have

$$(2.6) \qquad \left(\min_{i \leq i \leq m} a^i \cdot \ell\right)\left(\min_{i \leq j \leq r} b^j \cdot w\right) \leq \ell \cdot w.$$

In the next section we shall examine the analogs of (2.5) and (2.6) for a nonnegative matrix A and the blocking matrix B of Theorem 2.1.

3. The Length-Width Inequality and Max-Flow Min-Cut Equality.

Let A and B be m by n and r by n proper matrices having rows $a^1, \ldots, a^m$ and $b^1, \ldots, b^r$ respectively. Say that the max-flow min-cut equality holds for the pair A, B (in this order) if and only if, for each $w \in R^n_+$, it is true that in the linear program

$$(3.1) \qquad yA \leq w, \quad y \geq 0, \quad \max 1 \cdot y,$$

we have

$$(3.2) \qquad \max 1 \cdot y = \min_{1 \leq j \leq r} b^j \cdot w.$$

Similarly, say that the <u>length-width inequality</u> holds for A and B if and only if, for every $\ell \in R_+^n$, $w \in R_+^n$, we have

$$(3.3) \qquad \left(\min_{1 \le i \le m} a^i \cdot \ell \right) \left(\min_{1 \le j \le r} b^j \cdot w \right) \le \ell \cdot w .$$

<u>Theorem 3.1.</u> (i) Let A and B be a pair of blocking matrices. Then the max-flow min-cut equality holds for A, B (in either order) and the length-width inequality holds for A, B.

(ii) Let A and B be proper matrices whose rows $a^1, \ldots, a^m$ and $b^1, \ldots, b^r$ satisfy $a^i \cdot b^j \ge 1$. If the length-width inequality holds for A, B, then A and B are a blocking pair.

(iii) Let A and B be proper matrices. If the max-flow min-cut equality holds for A, B (in that order), then A and B are a blocking pair.

<u>Proof.</u> (i) Suppose that A and B are a blocking pair. The max-flow min-cut equality follows from Theorem 2.1 and the duality theorem for linear programs. The linear program dual to (3.1) is

$$(3.4) \qquad \begin{aligned} Ab &\ge 1 , \\ b &\ge 0 , \\ \min\ & w \cdot b . \end{aligned}$$

The minimum in this program is achieved at an extreme point of the constraint set $\mathcal{B} = \{b \in R_+^n \mid Ab \ge 1\}$, that is, by Theorem 2.1, at a row of B, and hence (3.2) holds.

To see that the length-width inequality holds, let

$$(3.5) \qquad \lambda = \min_{1 \le i \le m} a^i \cdot \ell = \min_{a \in \mathcal{A}} a \cdot \ell ,$$

$$(3.6) \qquad \omega = \min_{1 \le j \le m} b^j \cdot w = \min_{b \in \mathcal{B}} b \cdot w .$$

Here $a^1, \ldots, a^m$ are the rows of A, $b^1, \ldots, b^r$ are the rows of B, and $\mathcal{A} = \{a \in R^n_+ \mid Ba \geq 1\}$, $\mathcal{B} = \{b \in R^n_+ \mid Ab \geq 1\}$. The second equality in (3.5) (the second equality in (3.6)) follows from Theorem 2.1 and the fact that the minimum value of a nonnegative linear form defined over $\mathcal{A}(\mathcal{B})$ occurs at an extreme point of $\mathcal{A}(\mathcal{B})$.

If either $\lambda = 0$ or $\omega = 0$, the length-width inequality holds trivially. If both $\lambda \neq 0$, $\omega \neq 0$, then we have $(\ell/\lambda) \cdot a \geq 1$ for all $a \in \mathcal{A}$ and $(w/\omega) \cdot b \geq 1$ for all $b \in \mathcal{B}$ by (3.5) and (3.6). Consequently $\ell/\lambda \in \hat{\mathcal{A}}$ and $w/\omega \in \hat{\mathcal{B}} = \mathcal{A}$. Thus $(\ell/\lambda) \cdot (w/\omega) \geq 1$, $\ell \cdot w \geq \lambda\omega$.

(ii) Let A and B be proper matrices whose rows satisfy $a^i \cdot b^j \geq 1$, and define $\mathcal{B} = \{x \in R^n_+ \mid Ax > 1\}$, $\mathcal{A} = \{x \in R^n_+ \mid Bx \geq 1\}$. Thus $\hat{\mathcal{B}} = \text{conv. hull}(\{a^1, \ldots, a^m\}) + R^n_+$ and $\hat{\mathcal{A}} = \text{conv. hull}(\{b^1, \ldots, b^r\}) + R^n_+$ satisfy $\hat{\mathcal{A}} \cdot \hat{\mathcal{B}} \geq 1$. Hence $\hat{\mathcal{A}} \subseteq \hat{\hat{\mathcal{B}}} = \mathcal{B}$. Assume now that the length-width inequality holds for A, B, and let $b \in \mathcal{B}$. We want to show that $b \cdot \mathcal{A} \geq 1$. Thus let $a \in \mathcal{A}$. By the length-width inequality applied to a, b, we have

$$a \cdot b \geq \left(\min_{1 \leq j \leq r} a \cdot b^j\right)\left(\min_{1 \leq i \leq m} b \cdot a^i\right) \geq 1,$$

since $a \in \mathcal{A}$, $b \in \mathcal{B}$. Thus $\mathcal{B} \subseteq \hat{\mathcal{A}}$, and hence $\hat{\mathcal{A}} = \mathcal{B}$.

(iii) Let A and B be proper matrices and assume that the max-flow min-cut equality holds for A, B. Let $a^1, \ldots, a^m$ be the rows of A, let $b^1, \ldots, b^r$ be the rows of B, and let $\mathcal{B} = \{x \in R^n_+ \mid Ax \geq 1\}$. Suppose that $b^k \notin \mathcal{B}$. By the separating hyperplane theorem, there is a $w \in R^n_+$ and an $\alpha > 0$ such that $w \cdot b^k < \alpha \leq w \cdot b$ for all $b \in \mathcal{B}$. But by the duality theorem for linear programs and the max-flow min-cut equality, we have $\min_{b \in \mathcal{B}} b \cdot w = \min_{1 \leq j \leq r} b^j \cdot w$, a contradiction. Hence $b^j \in \mathcal{B}$ for $1 \leq j \leq r$ and consequently $a^i \cdot b^j \geq 1$ for $1 \leq i \leq m$, $1 \leq j \leq r$. We shall finish the proof by showing that the length-width inequality holds. Thus let $\ell \in R^n_+$, $w \in R^n_+$, and define

$$\lambda = \min_{1 \leq i \leq m} a^i \cdot \ell, \quad \omega = \min_{1 \leq j \leq r} b^j \cdot w.$$

By the max-flow min-cut equality, there is a $y = (\eta_1, \ldots, \eta_m) \geq 0$ such that $yA \leq w$ and $1 \cdot y = \omega$. Thus

$$\lambda\omega = \lambda(1 \cdot y) = \lambda \sum_{i=1}^{m} \eta_i \leq \sum_{i=1}^{m} (a^i \cdot \ell)\eta_i = \ell \cdot \sum_{i=1}^{m} \eta_i a^i \leq \ell \cdot w.$$

Hence the length-width inequality holds for A, B, and thus A, B are a blocking pair.

Theorem 3.1 is sometimes useful in proving that two matrices A and B are a blocking apir. In Sec. 6, for example, we shall take A to be the incidence matrix of the clutter of permutation matrices and use Theorem 3.1 (iii) to pin down the blocking matrix B. Some other examples of this kind will also be discussed.

4. Contractions, Deletions, Paintings.

One can define operations of "contracting a coordinate" in a proper matrix A (or on the polyhedron $\mathcal{B} = \{b \in R_+^n \mid Ab \geq 1\}$) that are analogous to the operations of "contracting an element" or "deleting an element" in a graph, a matroid [13], or a clutter [10]. Just as for graphs, matroids, or clutters, these operations commute. Moreover, contracting the i-th coordinate in A corresponds to deleting the i-th coordinate in its blocking matrix B: the resulting matrices again constitute a blocking pair.

Let A be an m by n proper matrix. By a contraction of coordinate $i \in \{1, \ldots, n\}$ in A, we mean the following: drop the i-th column of A, and then drop all inessential rows in the resulting matrix. A deletion of coordinate i in A is the following: drop the i-th column of A, and then drop all rows that had a positive entry in column i. The new matrix obtained in each case is proper.

Geometrically, contracting coordinate i in A is an intersection of the polyhedron $\mathcal{B} = \{b \in R_+^n \mid Ab \geq 1\}$ with the hyperplane $\xi_i = 0$; deleting coordinate i is a projection of $\mathcal{B}$ on the hyperplane $\xi_i = 0$. It is easy to see that first contracting coordinate i, then deleting coordinate j, is equivalent to first deleting coordinate j, then contracting coordinate i. Thus one can unambiguously define "minors" of A (or of $\mathcal{B}$), just as in matroid theory [13], that arise by contracting some subset of coordinates and deleting some other subset, since the order in which operations are carried out is immaterial.

It is also not hard to see that if A , B are a blocking pair, and if coordinate i is contracted in A , then the blocker of the resulting matrix is obtained from B by coordinate i .

If A and b(A) are n-columned incidence matrices of a blocking pair of clutters, and if we partition the set $\{1, 2, \ldots, n\}$ into two sets, the "blue" set and the "red" set, say, then it is true that precisely one of the following alternatives holds: (a) there is a row of A all of whose 1's lie in the blue set; (b) there is a row of b(A) all of whose 1's lie in the red set. This "painting theorem" in fact characterizes the blocking relation for a pair of clutters on $\{1, 2, \ldots, n\}$ [6, 9]. The analogous painting theorem is valid also for blocking matrices A and B: For any partition of the column set $\{1, 2, \ldots, n\}$ into two sets, blue and red (empty sets not being excluded), there is either a row vector of A whose support is blue or a row vector of B whose support is red, but not both. (Here the <u>support</u> of a vector $a = (\alpha_1, \ldots, \alpha_n)$ is the set of $i \in \{1, 2, \ldots, n\}$ such that $\alpha_i \neq 0$.) It is clear that both alternatives cannot hold, since otherwise some row of A and some row of B would have inner product zero. That one of the alternatives must hold can be seen in various ways, e.g., assume there is no row of A whose support is blue, and consider the effect of deleting all red coordinates in A . Or consider the max-flow min-cut equality for A, B where the vector w has red coordinates zero and blue coordinates one.

In terms of the blocking polyhedra $\mathcal{B} = \{b \in R^n_+ \mid Ab \geq 1\}$ and $\mathcal{A} = \{a \in R^n_+ \mid Ba \geq 1\}$ defined by blocking matrices A and B , the painting theorem asserts that for any blue-red partition of the coordinates, there is either an $a \in \mathcal{A}$ with blue support or a $b \in \mathcal{B}$ with red support, but not both.

5. Blocking Polyhedra From Orthogonal Complements.

A particular class of blocking polyhedra can be generated in the following way. Let $\mathfrak{R}$ and $\mathfrak{R}^{\perp}$ be complementary orthogonal subspaces of R^n, and let $\mathcal{E}$ and $\mathcal{E}^{\perp}$ be the set of all elementary vectors of $\mathfrak{R}$ and $\mathfrak{R}^{\perp}$, respectively. (Here a vector of space $\mathfrak{R}$ is elementary if it is nonzero and has minimal support over all nonzero vectors of $\mathfrak{R}$[13]). Define

(5.1) $$\mathcal{E}_1 = \{a = (\alpha_1, \ldots, \alpha_n) \in \mathcal{E} \mid \alpha_1 = 1\},$$

(5.2) $$\mathcal{E}^{\perp} = \{b = (\beta_1, \ldots, \beta_n) \in \mathcal{E}^{\perp} \mid \beta_1 = 1\}.$$

The sets $\mathcal{E}_1$ and $\mathcal{E}_1^{\perp}$ are finite, say

(5.3) $$\mathcal{E}_1 = \{a^1, a^2, \ldots, a^m\},$$

(5.4) $$\mathcal{E}_1^{\perp} = \{b^1, b^2, \ldots, b^r\}.$$

For each $a^i = (1, \alpha_2^i, \ldots, \alpha_n^i) \in \mathcal{E}_1$, $b^j = (1, \beta_2^j, \ldots, \beta_n^j) \in \mathcal{E}_1^{\perp}$, let

(5.5) $$\bar{a}^i = (|\alpha_2^i|, \ldots, |\alpha_n^i|),$$

(5.6) $$\bar{b}^j = (|\beta_2^j|, \ldots, |\beta_n^j|),$$

and let A and B be the nonnegative matrices with n-1 colums having rows $\bar{a}^1, \ldots, \bar{a}^m$ and $\bar{b}^1, \ldots, \bar{b}^r$, respectively. It is easy to check that A and B are proper. We assert that A and B constitute a blocking pair of matrices.

This does not seem to be obvious, although one can see quickly that $\bar{a}^i \cdot \bar{b}^j \geq 1$, since $a^i \cdot b^j = 0$. That matrices A and B are a blocking pair follows from Theorem 3.1 (iii) and the proof in [9] that the max-flow min-cut equality holds for A, B.

One case of particular interest for combinatorics is where the space $\mathfrak{R}$(and hence $\mathfrak{R}^{\perp}$) is regular, i.e., can be viewed as the row space of a totally unimodular matrix [13]. (A matrix is totally unimodular if all its square submatrices have determinant 0 or ± 1.) In this case each elementary vector of $\mathfrak{R}$ (and of $\mathfrak{R}^{\perp}$) can be taken to have coordinates 0, 1, or -1, and consequently matrices A and B above are (0, 1) -matrices. For example, if the space $\Re$ is the row space of the (0, $\pm$ 1) vertex-edge incidence matrix of an oriented graph on n edges, the construction above yields A as the (0, 1)-incidence matrix of all cuts separating the two end nodes of edge 1 in the underlying unoriented graph with edge 1 suppressed, and B as the incidence matrix of all paths joining these two nodes in the same graph.

6. Other Examples of Blocking Polyhedra.

In this concluding section we describe some other examples of blocking polyhedra that have combinatorial interest. In each of the examples, we start with a (0, 1)-matrix A which can be viewed as the incidence matrix of a clutter on a finite set, and examine the blocking matrix B. It is usually difficult to determine B, and we have not succeeded in doing this for certain clutters about which little is known, such as all minimal node-covers in an arbitrary graph, or all Hamiltonian tours in a complete graph.

Let A be the incidence matrix of all n by n permutation matrices. Thus A has n! rows and n^2 columns corresponding to pairs i, j, for $i, j \in \{1, 2, \ldots, n\}$. We assert that the blocking matrix B consists of the essential rows of the following matrix B^+. For each $I \subseteq \{1, 2, \ldots, n\}$, $J \subseteq \{1, 2, \ldots, n\}$ such that $s(I, J) = |I| + |J| - n > 0$, let $b(I, J)$ be the n^2-vector having coordinates $1/s(I, J)$ for $i \in I$, $j \in J$, zero otherwise, and let B^+ be the matrix consisting of all rows $b(I, J)$. (Some of the rows of B^+ are inessential, but not many. If $I = \{1, 2, \ldots, n\}$, and $J = \{j_1, \ldots, j_k\}$ is not a singleton, the row is inessential, being a convex combination of the rows $b(I, \{j_1\}), \ldots, b(I, \{j_k\})$, and similarly for I not a singleton, $J = \{1, 2, \ldots, n\}$. It can be shown that all other rows of B^+ are essential, however.) That the matrix B is the blocking matrix of A follows from Theorem 3.1 (iii) and results of [8], where the max-flow min-cut equality is proved for the matrices A and B^+, and hence for the proper matrices A and B. As shown in [8], there is an efficient algorithm for solving the max-flow min-cut problem for the matrices A and B, in this order, based on the maximum flow routine for (ordinary) flows in networks. That is, the maximizing vector y in (3.1) and the minimizing row of B in (3.2) can be calculated explicitly and efficiently. The max-flow min-cut equality of course holds in the reverse order B, A, but we know of no efficient algorithm for finding the maximizing vector y here. Finding the minimizing row of A is the well-known optimal assignment problem, for which efficient methods are known. It seems likely that there is an alternative approach to the optimal assignment problem based on the max-flow min-cut equality for B, A, i.e., based on the above characterization of the permutation matrices as the extreme points of $\mathcal{Q} = \{a \in R_+^{n^2} \mid Ba \geq 1\}$. For example, consider the 7 by 7 assignment problem with cost matrix w shown in Fig. 2 below. An optimal y weights two rows of B positively: y({6, 7}, {2, 3, 4, 5, 6, 7}) = 1, y({2, 4, 5, 6, 7}, {3, 5, 6, 7}) = 2.

	1	2	3	4	5	6	7
1	3	6	2	0	0	0	0*
2	8	0*	4	0	2	3	6
3	9	2	0*	8	5	3	0
4	2	0	4	4	1*	5	5
5	0	9	7	3	8	1*	9
6	0*	3	4	5	8	8	5
7	0	9	9	1*	7	6	3

Fig. 2

The discussion above can be generalized to the linear programming problem known as the transportation problem [7].

We turn next to an example of a different kind. Let A be the incidence matrix of all minimal node-covers in a graph. It would be interesting to know a characterization of the row vectors of A as the extreme points of a polyhedron. Here, in contrast with the example above, inequalities characterizing the convex hull are not known. Neither do we know inequalities characterizing the vector sum of the convex hull and the nonnegative orthant, i. e., the rows of the blocking matrix B. Some examples may indicate the difficulty of determining B. Suppose A is the incidence matrix of all minimal covers in the complete graph on n vertices. Thus A is n by n with zeros down the main diagonal and ones in all other positions. It is not difficult to see in this case that the matrix B has $2^n - (n+1)$ rows, one corresponding to each subset $I \subseteq \{1, 2, \ldots, n\}$ such that $|I| \geq 2$; specifically, the row b(I) has coordinates $1/(|I|-1)$ in positions corresponding to $i \in I$, zeros elsewhere. Thus here again we have the situation in which the nonzero elements of each row of B are all equal, just as for the permutations. But this situation is not typical for this problem. For instance, consider the graph of a "wheel" with an odd number of "spokes." To be specific, consider a 5-sided wheel, shown in Fig. 3. The matrices A and B are shown in Fig. 3 also. Note the first row of B.

Another clutter arising in graph theory that has been studied in depth by Edmonds [1, 2] is where A is the incidence matrix of all perfect matchings in a complete graph. Edmonds has characterized the convex hull of such a clutter, and has described an efficient algorithm for determining a minimum-weight perfect matching for an arbitrary weight function defined on the edges of the graph. What is the blocking matrix B for A? The blocking clutter b(A) is described in [6], the description being deduced from Tutte's theorem characterizing graphs that contain a perfect matching [11]. All members of b(A) will yield rows of B, but what other kinds of rows does B have? It would appear that to answer this question, we need information about the maximum "number of disjoint matchings" contained in an arbitrary graph, at least in the sense of admitting rational weights on matchings, i. e., we need to know how to solve the max-flow problem (3. 1) for A and an arbitrary $w \geq 0$ defined on the edges of the graph. Good information on the (0, 1)-form of this problem (i. e., w a given (0, 1)-vector and y restricted to be a (0, 1)-vector) could well lead to a solution of the four-color problem. Even the rational form

A

1	2	3	4	5	6
0	1	1	1	1	1
1	1	1	0	1	0
1	0	1	1	0	1
1	1	0	1	1	0
1	0	1	0	1	1
1	1	0	1	0	1

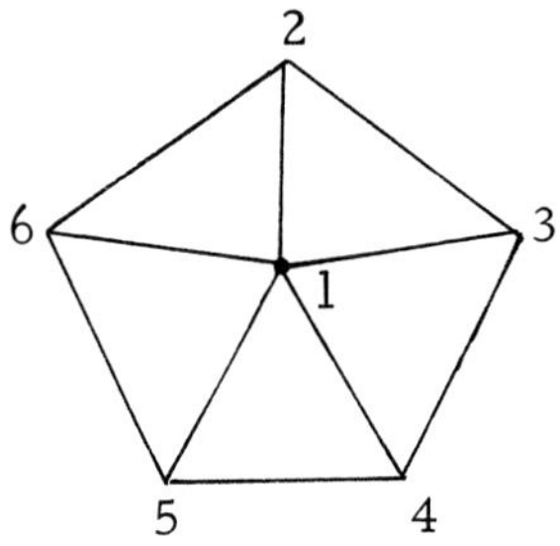

B

1	2	3	4	5	6
2/5	1/5	1/5	1/5	1/5	1/5
0	1/3	1/3	1/3	1/3	1/3
1/2	1/2	1/2	0	0	0
1/2	0	1/2	1/2	0	0
1/2	0	0	1/2	1/2	0
1/2	0	0	0	1/2	1/2
1/2	1/2	0	0	0	1/2
1	1	0	0	0	0
1	0	1	0	0	0
1	0	0	1	0	0
1	0	0	0	1	0
1	0	0	0	0	1
0	1	1	0	0	0
0	0	1	1	0	0
0	0	0	1	1	0
0	0	0	0	1	1
0	1	0	0	0	1

Fig. 3

of the problem appears to be unsolved, except in the bipartite case, where we are dealing with the permutation matrices.

In each of the examples discussed thus far in this section, the blocking matrix B for the incidence matrix A of a clutter appears to be more complicated in structure than a list of inequalities defining the convex hull of the clutter. We conclude with two examples in which this is not the case.

Let A be the incidence matrix of all spanning trees in a graph on edge-set $\{1, 2, \ldots, n\}$ (more generally, we could consider bases in a matroid). Here Edmonds has shown that the extreme points of the polyhedron

(6.1) $$\sum_{i=1}^{n} \xi_i = \text{rank}(\{1, 2, \ldots, n\}) ,$$

(6.2) $$\sum_{i \in I} \xi_i \leq \text{rank}(I) , \quad \text{all } I \subseteq \{1, 2, \ldots, n\} ,$$

(6.3) $$\xi_i \geq 0 ,$$

are precisely the rows of A[5]. (It is enough to consider sets $I \subseteq \{1, \ldots, n\}$ in (6.2) that are spans.) It can also be shown, using results of Tutte [12] and of Edmonds [3, 4], that the blocking matrix B of A consists of the essential rows of the matrix B^+ which has a row $b(\bar{I})$ corresponding to each nonempty complement $\bar{I}$ of a span: the row $b(\bar{I})$ has components $1/(\text{rank}(\{1, 2, \ldots, n\}) - \text{rank}(I))$ in positions corresponding to elements of $\bar{I}$, zeros elsewhere. Edmonds has described an efficient algorithm for finding the maximum number of edge-disjoint spanning trees in a graph [3, 4] (or the maximum number of disjoint matroid bases) and it is not hard to extend this to the case of an arbitrary weight function $w \geq 0$ defined on the edges. In other words, the max-flow problem (3.1) in either integer or rational form has been solved for this clutter A, just as it has for the clutter of permutation matrices. The max-flow problem in the other direction, that is, for B, A, has also been solved: Finding a min-cut is the well-known minimum spanning tree problem and it is not difficult to describe an algorithm for finding a corresponding maximizing vector y. Here one can make good use of contractions and deletions in determining a min-cut and a max-flow.

Our last example deals with edge-covers in a bipartite graph. Let A be the incidence matrix of all minimal covers of nodes by edges in a bipartite graph having r nodes in one part, s in the other, and n edges. (Since we want to consider an arbitrary weight function $w \geq 0$ defined on the edges, we could suppose without loss of generality that the graph is a complete bipartite graph having rs edges. Then A has rs columns and a row corresponding to every minimal cover, i. e., a row corresponding to every r by s (0, 1)-matrix having at least one 1 in each of its rows and columns, and which is minimal with respect to this property.) The blocking matrix B of A here is simply the node-edge incidence matrix of the graph. In other words, the incidence matrix b(A) of the blocking clutter is equal to B in this instance. This can be shown in various ways, perhaps the easiest of which is to start with the (0, 1)-matrix B and ask for its blocking matrix. It is well-known that the matrix B is totally unimodular, and it follows from this that all extreme points of $\mathcal{A} = \{a \in R_+^n \mid Ba \geq 1\}$ are (0, 1)-vectors. Consequently, the blocking matrix A of B is the one described above. In connection with this example, consider the max-flow min-cut equality for A, B (in this order). Here it can be shown that if $w \geq 0$ has integer coordinates, then there is a maximizing y in (3.1) having integer coordinates, which leads to the following theorem: The maximum number of edge-disjoint covers (of nodes by edges) in a bipartite graph is equal to the minimum valence in the graph. This is a companion to the well-known König theorem that a bipartite graph having maximum valence k can be decomposed into a sum of k matchings, i. e., the minimum number of colors required in an edge-coloring is equal to the maximum valence. In terms of (0, 1)-matrices, the König theorem says that if G is a given (0, 1)-matrix, then the least k for which we have

$$(6.4) \qquad G \leq M_1 + \ldots + M_k ,$$

where each M_i is a (0, 1)-matrix having at most one 1 in each row and column, is equal to the largest row or column sum of G. The companion theorem says that the largest k for which

$$(6.5) \qquad G \geq C_1 + \ldots + C_k ,$$

where each C_i is a (0, 1)-matrix having at least one 1 in each row and column, is equal to the smallest row or column sum of G .

We have said nothing about the length-width inequality in these examples. In the examples where the blocking matrix B is known (for permutations, trees, and edge-covers in a bipartite graph), the corresponding length-width inequality appears to be a new result in each case.

REFERENCES

1. Edmonds, J., "Paths, trees, and flowers," Canad. J. Math., 17 (1965), 449-467.

2. Edmonds, J., "Maximum matching and a polyhedron with 0, 1-vertices," J. Res. Natl. Bur. Standards Sect. B 69 (1965), 125-130.

3. Edmonds, J. , "Minimum partition of a matroid into independent sets," J. Res. Natl. Bur. Standards Sect. B (1965), 67-72.

4. Edmonds, J. , "Lehman's switching game and a theorem of Tutte and Nash-Williams," J. Res. Natl. Bur. Standards, Sect. B 69(1965), 73-77.

5. Edmonds, J. , "Matroids and the greedy algorithm," to appear (lecture, Princeton, August 1967).

6. Edmonds, J. and D. R. Fulkerson, "Bottleneck extrema", to appear in Journal Combinatorial Theory.

7. Ford, L. R., Jr. and D. R. Fulkerson, Flows in Networks, Princeton University Press (1962).

8. Fulkerson, D. R., "The maximum number of disjoint permutations contained in a matrix of zeros and ones," Canad. J. Math. 16 (1964), 729-735.

9. Fulkerson, D. R., "Networks, frames, blocking systems," Mathematics of the Decision Sciences, Lectures in Applied Mathematics, Vol. 11, Amer. Math. Soc. (1968).

10. Lehman, A., "On the width length inequality," (mimeo. 1965) to appear in SIAM Journal.

11. Tutte, W. T., "The factors of graphs," Canad. J. Math. 4, (1952), 314-329.

12. Tutte, W. T. , "On the problem of decomposing a graph into n connected factors," J. London Math. Soc. 36 (1961), 221-230.

13. Tutte, W. T., "Lectures on matroids," J. Res. Natl. Bur. Standards Sect. B. 69 (1965), 125-130.

This research is supported by the United States Air Force under Project RAND - Contract No. F44620-67-C-0045.

Connectivity in Matroids

W. T. TUTTE

Summary

This expository paper describes in general terms how a certain connectivity theorem for graphs is generalized as a theorem about matroids. For detailed proofs the reader is referred to a paper with the same title which appeared in the Canadian Journal of Mathematics in 1966 [2].

1. Matroids and graphs.

There are several equivalent definitions of a matroid. In this paper we use Whitney's definition in terms of "circuits", which runs as follows.

A matroid is defined by a finite set E of elements called "cells", and a family of non-null subsets of E called "circuits". The circuits are required to satisfy the following axioms.

(i) If X and Y are circuits, then X is not a proper subset of Y.

(ii) If X and Y are circuits, $a \in X \cap Y$ and $b \in X - Y$, then there is a circuit Z such that $b \in Z \subseteq (X \cup Y) - \{a\}$.

It is possible to obtain a matroid $P(G)$ from a graph G by taking the edges of G as the cells of $P(G)$, and the circuits of G, regarded as sets of edges, as the circuits of $P(G)$. The two axioms can be verified for $P(G)$ by elementary graph-theoretical methods. $P(G)$ is called the polygon-matroid of G in [2]. Not every matroid can be interpreted as the polygon-matroid of a graph.

One method of finding theorems about matroids is to

select a known theorem about graphs, state it as a proposition solely about about circuits and edges, and then try to prove this proposition as a theorem about general matroids (with the edges as cells). Often this procedure is successful, though sometimes the proposition for general matroids has to be weakened a little.

We proceed to discuss an example of this method. The graphic theorem is concerned with "3-connection". It is given in [1].

2. Connectivity in graphs and matroids.

For any graph G we denote the set of edges by E(G) and the set of vertices by V(G). For the cardinality of a set S we write $|S|$. A graph is called simple if it has no loop or multiple join, that is no circuit of fewer than 3 edges.

For a non-negative integer n the definition of n-connection given in [1] is as follows.

A graph is n-separated if it is the union of two subgraphs H and K with the following properties.

(i) $E(H) \cap E(K)$ is null.

(ii) $|V(H) \cap V(K)| \leq n$.

(iii) Each of the subgraphs H and K has a vertex not belonging to the other.

A graph that is not $(n - 1)$-separated is called n-connected. Equivalently we can say that a graph is n-connected if and only if it is not k-separated for any $k < n$.

According to this definition every graph is 0-connected. A graph is 1-connected if and only if it is connected, and 2-connected if and only if it is non-separable. A complete graph is n-connected for every n. There are of course other definitions of n-connection for which the last assertion is false.

Let G have an edge A that is not a loop. Then we define G'_A as the graph derived from G by deleting A, and G''_A as the graph derived from G by contracting A and its two ends into a single new vertex.

Suppose G is 3-connected. Then we call A an essential edge of G if neither G'_A nor G''_A is both simple and 3-connected. An example of a 3-connected graph in which every edge is essential is provided by the wheel of k spokes ($k \geq 3$). This graph is formed from a circuit of k vertices, the rim, by adjoining a new vertex called the hub and then joining the hub to each vertex of the rim by a single edge called a spoke. The main result of [1] is the following

THEOREM: Let G be a simple 3-connected graph, with at least four vertices, in which every edge is essential. Then G is a wheel.

It is this theorem that we propose to generalize to matroids. We recall that the first step should be to restate it in terms of circuits and edges. We therefore try to put the definition of n-connection into this form.

Some progress can be made by replacing Condition (iii) of the definition of n-separation by the following.

$$\text{(iii')} \qquad |E(H)| \geq n \leq |E(K)| \ .$$

We also require H and K to be non-null even when $n = 0$. This change of course gives us a new kind of n-separation, from which we can derive a new kind of n-connection.

It is easily seen that a graph which is n-separated according to the first definition is also k-separated, for some $k \leq n$, according to the second. Moreover if the graph is simple and $n \leq 2$ then the second kind of n-separation implies the first.

Proceeding from the new definition of n-separation we say that a graph is $(n+1)$-connected if it is not k-separated for any $k \leq n$. By the preceding paragraph the new definition of 3-connection is, for simple graphs, equivalent to the old.

We therefore replace (iii) by (iii') in the definition of n-separation.

Let us now define the connectivity $\lambda(G)$ of a graph

G as the greatest integer n such that G is n-connected. In some trivial cases such as the triangle no such integer exists, and we then write $\lambda(G) = \infty$. If G is connected we can say that $\lambda(G)$ is the least integer n such that some set of n vertices of G separates $E(G)$ into two sets each with at least n members (or is infinite if no such n exists). We shall in fact assume G to be connected from now on. This is not a very drastic assumption; any non-null graph can be made connected by suitable identifications of vertices, without changing the polygon-matroid.

We still have the mention of vertices in Condition (ii). We must get rid of this since matroids have no vertices. Some help can be obtained from the properties of $p_1(G)$, the Betti number of dimension 1 of a graph G . This can be defined as the least integer k such that the deletion of some k edges of G destroys every circuit. Such a number can equally well be defined for a matroid M . We can call it the <u>circuit-rank</u> of M . In [2] it is denoted by $r(M)$ and called simply the <u>rank</u> of M , though it is not the same as the Whitney rank.

For some connected graph G , let $E(G)$ be partitioned into two complementary subsets S and T . Let these subsets define subgraphs $G \cdot S$ and $G \cdot T$ of G respectively (without isolated vertices). Denote the number of common vertices of $G \cdot S$ and $G \cdot T$ by $\eta(G; S, T)$. A little elementary graph theory reveals that

$$p_1(G) - p_1(G \cdot S) - p_1(G \cdot T) + 1$$

$$= \eta(G; S, T) - p_0(G \cdot S) - p_0(G \cdot T) + 2$$

where $p_0(H)$ is the number of components of the graph H .

Now consider a matroid M on a set E of cells. If S is a subset of E then those circuits of M that are confined to S are the circuits of a matroid on S . We denote this matroid by $M \times S$. It is also known that there is a matroid $M \cdot S$ whose circuits are the minimal non-null intersections with S of the circuits of M . For a graph G it is found that the following identities hold.

$$P(G \cdot S) = P(G) \times S,$$

$$P(G'_A) = P(G) \times (E(G) - \{A\}),$$

$$P(G''_A) = P(G) \cdot (E(G) - \{A\}) .$$

If S and T are complementary subsets of E we write.

$$\xi(M; S, T) = r(M) - r(M \times S) - r(M \times T) + 1 .$$

We now have

$$\xi(P(G); S, T) = \eta(G; S, T) - p_0(G \cdot S) - p_0(G \cdot T) + 2 .$$

When $\lambda(G)$ is finite we can choose S and T so that $|S| \geq \lambda(G) \leq |T|$ and $\eta(G; S, T) = \lambda(G)$. It is then easy to show that $G \cdot S$ and $G \cdot T$ are connected. It follows that $\xi(P(G); S, T) = \lambda(G)$.

At this stage it may occur to us that perhaps $\xi(P(G); S, T)$ can be used instead of $\eta(G; S, T)$ in the definition of $\lambda(G)$. To assist in the discussion of this hypothesis we define the connectivity $\lambda(M)$ of a matroid M as follows. If there is an integer k such that $\xi(M; S, T) = k$ for some S and T satisfying $|S| \geq k \leq |T|$, then $\lambda(M)$ is the least such integer. If there is no such integer k then $\lambda(M)$ is infinite. The results we have just obtained can now be written as

$$\lambda(P(G)) \leq \lambda(G) .$$

The reverse of this inequality is valid but is much more difficult to prove. In [2] the following theorem is established, but the proof is surprisingly complicated.

Let G be a connected graph. Let S and T be complementary subsets of $E(G)$ such that

$$\eta(G;\ S,\ T) \leq k + p_0(G \cdot S) + p_0(G \cdot T) - 2$$

and

$$|S| \geq k \leq |T|,$$

where k is a positive integer. Then $\lambda(G) \leq k$.

If $\lambda(P(G))$ is finite we can choose S and T so that $\xi(P(G);\ S,\ T) = \lambda(P(G))$ and $|S| \geq \lambda)P(G)) \leq |T|$. We note that $\xi(P(G);\ S,\ T)$ is positive by definition. Hence by one of our earlier formulae the conditions of the above theorem hold with $k = \lambda(P(G))$. We deduce that $\lambda(G) \leq \lambda(P(G))$.

We observe that $\lambda(G)$ and $\lambda(P(G))$ are equal, for every connected graph G. We therefore regard the connectivity of a matroid, as defined above, as an appropriate generalization of the notion of the connectivity of a graph.

3. Wheels and whirls.

Having defined the connectivity of a matroid we may seek an analogue of the wheel theorem. Such an analogue is exhibited in [2]. The proof is much longer than that for the graphic theorem. It uses over and over again, the auxiliary theorem

$$\xi(M;\ S,\ T) + \xi(M;\ U,\ V)$$

$$\geq \xi(M;\ S \cap U,\ T \cup V) + \xi(M;\ S \cup U,\ T \cap V),$$

which can be proved by elementary matroid theory.

An essential cell of a 3-connected matroid M is defined by analogy with an essential edge of a 3-connected graph. A cell A of M is essential if neither $M \times (E - \{A\})$ nor $M \cdot (E - \{A\})$ is 3-connected. The 3-connected matroids in which every cell is essential are finally classified as the "wheels" and the "whirls". A matroidal wheel is naturally the polygon-matroid of a graphic wheel. A whirl is derived from a wheel of the same number of cells by the following operations. All the circuits of the wheel are retained with the exception of the rim R. Instead of R we recognize as circuits the sets formed from R by adjoining a single spoke.

The main theorem of [2] is as follows.

THEOREM: A 3-connected matroid has all its cells essential if and only if it is a wheel or a whirl.

REFERENCES

1. W. T. Tutte. A theory of 3-connected graphs, Indag. Math., 23, 1961, 441-455.

2. W. T. Tutte. Connectivity in matroids, Canad. J. Math., 18, 1966, 1301-1324.

The Use of Circuit Codes in Analog-to-Digital Conversion

VICTOR KLEE

The graph of the d-dimensional cube is here denoted by I(d). Thus the 2^d vertices of I(d) are the various d-tuples of binary digits, two vertices being joined by an edge if and only if they differ in precisely one coordinate. A d-dimensional circuit code of spread s is a (simple) circuit C in I(d) such that for any two vertices x and y of C,

$$\delta_{I(d)}(x, y) < s \implies \delta_C(x, y) = \delta_{I(d)}(x, y).$$

Here δ is the usual combinatorial distance of graph theory, the distance between two vertices of a graph being the minimum number of edges that form a path joining the two vertices. Thus every circuit in I(d) is of spread 1, and a circuit C is of spread 2 if and only if every edge of I(d) joining two vertices of C is itself an edge of C. More picturesquely, a circuit C in I(d) is of spread s provided that a runner racing around C cannot find a shortcut between two vertices of C consisting of fewer than s edges of I(d).

The circuits of length 2^d in I(d) (the Hamiltonian circuits) are commonly called Gray codes and are used to reduce the effects of the inevitable quantization error in various analog-to-digital conversion systems. However, no such circuit is of spread 2 and its use in conversion systems may consequently result in another sort of error caused by malfunctioning of the equipment. For partial control of such errors, circuit codes of spread 2 were introduced by Kautz [14] in 1958 under the name of unit-distance error-checking codes or snake-in-the-box codes. Circuit codes of spread $s > 2$, which have additional error-checking properties, have also been called circuit codes of minimum distance s and SIB_s codes.

The present brief expository account of circuit codes consists of three parts in addition to this introduction. The first part describes the usefulness of circuit codes in a typical analog-to-digital conversion system. The second summarizes the known results on $C(d, s)$, the maximum length of d-dimensional circuit codes of spread s. The third suggests avenues of future research on circuit codes.

1. Circuit codes in analog-to-digital conversion.

Consider the problem of recording, by means of ordered d-tuples of binary digits, the positions of a rotating wheel. The wheel is divided by means of k radii into k congruent sectors, and some d-tuple is to be associated with each sector. To avoid ambiguity, no two sectors should have the same d-tuple; thus $2^d \geq k$. The wheel is further divided by means of $d - 1$ smaller circles concentric with it, so that each sector consists of a sequence of d regions of the wheel, progressing outward from the center. Some of the regions are punched out to permit the passage of light and the wheel's position is read by a reader consisting of a linear array of d photoelectric cells aligned with a radius of the wheel. The reader is fixed in position, so that various d-tuples of binary digits are recorded as the wheel turns. For example, a reading of d zeros indicates that none of the regions is punched in the sector aligned with the reader, while a reading of d ones indicates that all are punched. Now, what is a reasonable way of encoding the wheel - that is, of assigning k different d-tuples of binary digits to the various sectors of the wheel?

Let us first consider the case in which $d = 3$ and $k = 8$, so that all triples of binary digits are to be used. Using them in their "natural" order would result in the coding shown in Figure 1, where the shaded regions correspond to zeros (unpunched) and the unshaded regions to ones (punched). However, this is an unusually bad coding because of the inevitable quantization problems associated with

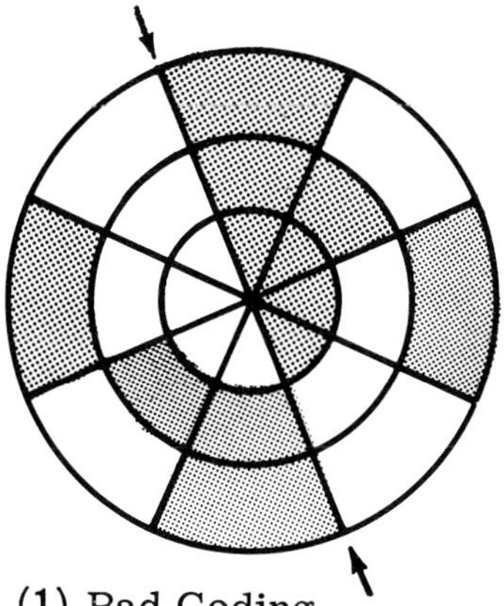

(1) Bad Coding

analog-to-digital conversion. When the reader is nearly aligned with a radius separating two sectors, the jth photoelectric cell may take its reading from the jth region of one sector or from the jth region of the other sector. Hence the jth digit of the reading is meaningless when the jth region of one sector is punched and that of the other is not. In Figure 1, for example, any triple may be recorded when the reader is nearly aligned with either of the two radii indicated by an arrow.

Though the quantization error cannot be entirely eliminated, its effect may be minimized by using a coding (such as the one shown in Figure 2) in which the d-tuples assigned to adjacent sectors differ in only one coordinate. With such a coding, there are only two readings that may be obtained when the reader is nearly aligned with the radius separating two sectors. As each of these readings is associated with one of the sectors, the analog-to-digital conversion is made as accurately as it can be for the given number of sectors, assuming the equipment functions properly. The requirement that d-tuples assigned to adjacent sectors differ in only one coordinate is equivalent to the condition that the successive d-tuples are the vertices of a circuit in I(d). (The second part of Figure 2 depicts the circuit in I(3) that corresponds to the coding of the first part.) In a complete

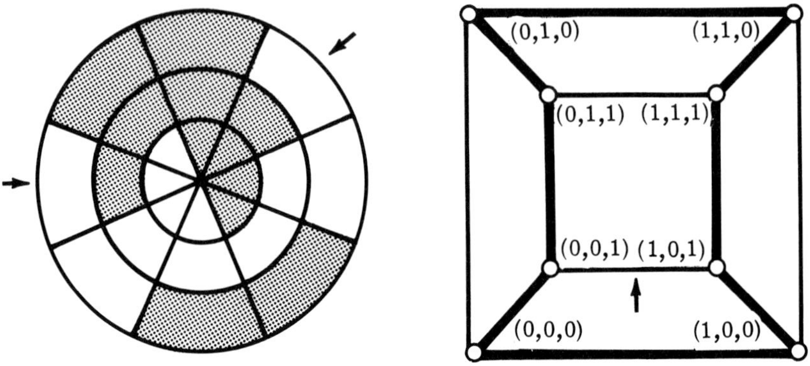

(2) Gray Code

traversal of any circuit of I(d), each coordinate changes an even number of times and hence the circuit's length is even. Thus the number k of sectors is henceforth assumed to be even. The circuits for which $k = 2^d$ are commonly called <u>Gray codes</u>.

In addition to unavoidable quantization errors, errors may result from malfunctioning of the equipment. In Figure 2, for example, failure of the reader's first cell to fire when it should could result in a reading of 001 rather than 101 (note arrows in both parts of Figure 2). Any vertex v of I(d) is adjacent to d others, while in any given circuit C through v there are only two vertices C-adjacent to v. Hence for any vertex of a d-dimensional Gray code there are d-2 single-digit errors which, like the one just described, are undetectable and result in readings that indicate sectors not even adjacent to the correct one. The effect of single-digit errors may be minimized by a coding that omits some of the vertices of I(d) and uses a circuit code C of spread 2 rather than a Gray code. When the circuit is of spread 2 any single-digit error results in a reading that either (a) indicates a sector adjacent to the correct one or (b) does not correspond to any of the code points. Any error of type (a) is undetectable but can be tolerated as its magnitude is small (especially when k is large). Errors of type (b) are readily detectable if the set of code points admits a simple description, and the resultant readings are simply ignored. An error of type (b) is in general not correctable, as it may correspond to a vertex v of I(d) ~ C that is adjacent to several vertices of C .

The change from a Gray code to a circuit code C of spread 2 necessarily involves either (1) a dimunition of k and hence of the potential accuracy (neglecting malfunctions) of the analog-to-digital conversion (see Figure 3), or (2) an increase of d , which is likely to result in increased complexity of the circuitry involved (see Figure 4). The choice

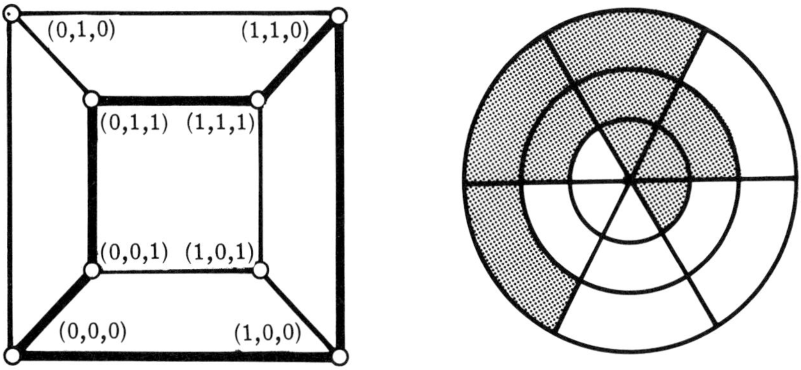

(3) Circuit code of spread 2 and dimension 3

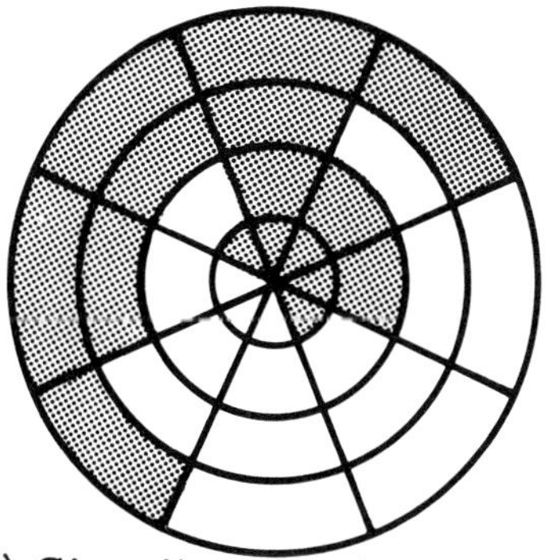

(4) Circuit code of spread 2 and dimension 4

between (1) and (2) must be considered whenever a circuit code is replaced by another one of greater spread.

When a circuit code C is of spread $s > 2$ any r-digit error, with $r < s$, results in a reading that either (a) indicates a sector precisely r sectors removed from the correct one or (b) does not correspond to any of the code points. An error of type (a) is undetectable but it may be tolerable. An error of type (b) is readily detectable if the set of code points admits a simple description. Further, an error of type (b) is "approximately correctable" when $2r < s$, for if x is the correct vertex of C, v is the erroneously recorded vertex of $I(d) \sim C$, and y is any vertex of C such that $\delta_{I(d)}(y, v) \leq r$, then $\delta_C(x, y) \leq 2r$.

2. Known results on $C(d, s)$.

There are three parameters associated with each circuit code: its dimension d, spread s, and length ℓ. It is natural to attempt:

for given d and s, to maximize ℓ, as that maximized the potential accuracy of the analog-to-digital conversion;

for given d and ℓ, to maximize s, as that maximizes the error-checking properties of the code;

for given s and ℓ, to minimize d, as that is related to minimizing the complexity of the required circuitry. The existing literature deals almost entirely with the first problem, though of course the three problems are closely related.

For all d and s, let $C(d, s)$ denote the maximum length of d-dimensional circuit codes of spread s. Then $C(d, 1) = 2^d$ for all $d \geq 2$; that is Gray codes exist in all dimensions ≥ 2. Among the many references in which Gray codes are discussed, we mention Keister, Ritchie and Washburn [15, chap. 11], Gray, Levonian, and Rubinoff [13], Caldwell [2, pp. 391-396], and Fifer [9, pp. 308-311] for the use of such codes in analog-to-digital conversion, Fischmann [10], Cohn and Even [4], and the Russian references of the latter for the use of Gray code counters in purely digital systems, and Gilbert [11], Mills [19] and Abbott [1] for more mathematical studies of Gray codes.

For $s > 1$, it is only when d/s is small that the exact value of $C(d, s)$ is known. Specifically,

$$C(d, s) = 2d \quad \text{for} \quad d < \lceil 3s/2 \rceil + 2;$$

$$C(\lceil 3s/2 \rceil + 2, s) = 4s + 4 \quad \text{for odd} \quad s;$$

$$C([3s/2] + 2, s) = 4s + 6 \quad \text{for even} \quad s;$$

$$C([3s/2] + 3, s) = 4s + 8 \quad \text{for odd} \quad s \geq 9.$$

These results are due to Singleton [20] and Douglas [7]. In addition, it is known that $C(6, 2) = 26$ (Davies [6]), $C(7, 3) = 24$, $C(10, 5) = 28$ or 30, and $C(13, 7) = 36$ or 38 (Douglas [7]).

The best upper bounds on $C(d, s)$ are those of Douglas [8], who proved

$$C(d, 2) \leq 2^{d-1} - \frac{2^d - 12}{7d(d-1)^2 + 2},$$

and Chien, Freiman, and Tang [3], who proved for $t \geq 1$ that

$$C(d, 2s+1) \leq \frac{2^d}{\binom{d}{t} - 2\binom{d-1}{t-1}}$$

$$\text{for} \quad d > 2t + 1.$$

Chien, Freiman, and Tang [3] also obtained an upper bound for $C(d, 2s + 2)$ that is somewhat smaller than the above bound for $C(d, 2t + 1)$, and their bound was improved by Douglas [8].

The best lower bound for $C(d, 2)$ is due to Danzer and Klee [5] who showed

$$C(d, 2) \geq \frac{7}{4} \frac{2^d}{d-1} \quad \text{for} \quad d \geq 5.$$

The best lower bounds for $C(d, s)$ when $s > 2$ are due to Singleton [20] and Klee [16]. It follows from their results that

$$\left.\begin{array}{ll} & C(d,3) \\ & C(d,4) \\ \\ (\text{odd } s) & C(d,s) \\ (\text{even } s) & C(d,s) \end{array}\right\} \succ \mu^d \text{ for all positive } \mu < \left\{\begin{array}{l} \sqrt[3]{3} \\ \sqrt[3]{3} \\ \\ \sqrt[s+1]{4} \\ \sqrt[s]{4} \end{array}\right.$$

where $C(d, s) \succ \mu^d$ means $\lim_{d\to\infty} C(d, s)/\mu^d = \infty$.

For references to other studies of $C(d, s)$ whose results have been superseded by those mentioned above, see the expositions of Klee [17], [18]; see also Glagolev [12].

3. Directions for future research.

In addition to the obvious problem of improving the partial results described in the preceding section, we suggest six more specific directions for future research on circuit codes.

(1) The lower bounds of Danzer and Klee [5] and Klee [16] depend on a construction whereby, roughly speaking, a long d-dimensional code of spread s is obtained by combining a long $(d/2)$-dimensional code of spread s and a long $(d/2)$-dimensional code of spread $s-1$. However, this construction seems to work only when s is even. Discovery of an analogous procedure for odd s would result in considerable improvement in lower bounds for $C(d, s)$ for all $s > 2$. (For a comment on this, see [16, pp. 524-525].)

(2) The reader familiar with the "classical" error-correcting codes will recognize the close analogy between the considerations encountered there and those in our Section 1. Note also that any circuit code of spread s and length ℓ must contain s pairwise disjoint error-correcting codes of minimum distance s and cardinality $[\ell/s]$. Another connection between circuit codes and error-correcting codes was exploited in Vasil'ev's construction [21] of long circuit codes of spread 2. However, since error-correcting codes have been explored so much more extensively than circuit codes, the connections between the two can probably be exploited to yield new results on circuit codes.

(3) In addition to constructing circuit codes of maximum length ℓ for given d and s, there is interest in procedures for constructing those of specified length. For example, there exists an 11-dimensional circuit code of spread 2 whose length is 416, but for recording angles with an accuracy of one degree it is more convenient to use a code whose length is exactly 360. Klee [16] has described a procedure for constructing codes of specified length, but more should be done in this direction.

(4) The basic problem of decoding seems not to have been seriously considered except in the case of Gray codes. When $s = 2$, the problem is that of producing circuit codes of spread 2 whose vertex-sets can be described as simply as possible. For $s > 2$ there are two types of decoding problems, according to whether or not one wants to make the sort of "approximate corrections" mentioned in Section 1.

(5) Except in the case of Gray codes, little attention has been given to the complexity of the circuitry required for the actual use of various circuit codes in analog-to-digital conversion. Most of the existing long circuit codes have symmetry properties that should be helpful in simplifying the circuitry, but no detailed study of this matter has been made.

(6) The definition of the spread of a circuit code makes no distinction between errors that lead to a 0 reading when 1 is correct and those that lead to a 1 reading when 0 is correct. However, in existing analog-to-digital conversion systems, one type of error may be considerably more probable than the other. There are several ways in which this asymmetry might be taken into account in designing circuit codes for use in conversion systems.

REFERENCES

1. H. L. Abbott, Hamiltonian circuits and paths on the n-cube, Canad. Math. Bull. 9 (1966), 557-562.

2. S. H. Caldwell, Switching circuits and logical design, John Wiley and Sons, Inc., New York, 1958.

3. R. T. Chien, C. V. Freiman, and D. T. Tang, Error correction and circuits on the n-cube. Proc. Second Allerton Conference on Circuit and System Theory, Sept. 28-30, 1964, Univ. of Illinois, Monticello, Ill., pp. 899-912.

4. M. Cohn and S. Even, A Gray code counter, IEEE Trans. Computers C-18 (1969), 662-664.

5. L. Danzer and V. Klee, Lengths of snakes in boxes, J. Combinatorial Theory 2 (1967), 258-265.

6. D. W. Davies, Longest 'separated' paths and loops in an N cube, IEEE Trans. Electronic Computers EC-14 (1965), 261.

7. R. J. Douglas, Some results on the maximum length of circuits of spread k in the d-cube, J. Combinatorial Theory 6 (1969), 323-339.

8. ____________, Upper bounds on the length of circuits of even spread in the d-cube, J. Combinatorial Theory 7 (1969), 206-214.

9. S. Fifer, Analogue Computation, vol. 2, McGraw Hill Book Company, New York, 1961.

10. A. F. Fischmann, A Gray code counter, IRE Trans. Electronic Computers EC-6 (1957), 120.

11. E. N. Gilbert, Gray codes and paths on the n-cube, Bell System Tech. J. 37 (1958), 815-826.

12. V. V. Glagolev, An upper estimate of the length of a cycle in the n-dimensional unit cube (Russian), Diskret. Analiz., No. 6 (1966), 3-7.

13. H. J. Gray, Jr., P. V. Levonian, and M. Rubinoff, An analogue-to-digital converter for serial computing machines, Proc. IRE 41 (1953), 1462-1465.

14. W. H. Kautz, Unit-distance error-checking codes, IRE Trans. Electronic Computers EC-7 (1958), 179-180.

15. W. Keister, A. E. Ritchie, and S. H. Washburn, The Design of Switching Circuits, D. Van Nostrand Co., Inc., New York, 1951.

16. V. Klee, A method for constructing circuit codes, J. Assoc. Comput. Mach. 14 (1967), 520-528.

17. ______, Long paths and circuits on polytopes, Chap. 17 in B. Grünbaum's Convex Polytopes, Interscience-Wiley, London-New York-Sydney 1967.

18. ______, What is the maximum length of a d-dimensional snake?, Amer. Math. Monthly 77 (1970), 63-65.

19. W. H. Mills, Some complete cycles on the n-cube, Proc. Amer. Math. Soc. 14 (1963), 640-643.

20. R. C. Singleton, Generalized snake-in-the-box codes, IEEE Trans. Electronic Computers EC-15 (1966), 596-602.

21. Ju L. Vasil'ev, On the length of a cycle in the n-dimensional unit cube, Soviet Math. Dokl. 4 (1963), 160-163 (transl. from Dokl. Akad. Nauk SSSR 148 (1963), 753-756).

Work supported in part by Boeing Scientific Research Laboratories and the Office of Naval Research.

Some Mapping Problems for Tournaments

J. W. MOON

1. Introduction.

A (round-robin) tournament T consists of a set of nodes 1, 2, ..., n such that each pair of distinct nodes i and j is joined by exactly one of the oriented arcs $\overrightarrow{ij}$ or $\overrightarrow{ji}$. If the arc $\overrightarrow{ij}$ is in T we say i dominates j and write $i \rightarrow j$; more generally, if every node of a subtournament Y dominates every node of a subtournament Z we say Y dominates Z and write $Y \rightarrow Z$. If i is a node of a tournament T then $\Gamma(i)$ denotes the set of nodes dominated by i; the score of i is the number s_i of nodes in $\Gamma(i)$. Two tournaments are isomorphic, or anti-isomorphic, if there exists a one-to-one mapping between their nodes that preserves, or reverses, all dominance relations between nodes. The non-isomorphic tournaments with three and four nodes are illustrated in Figure 1.

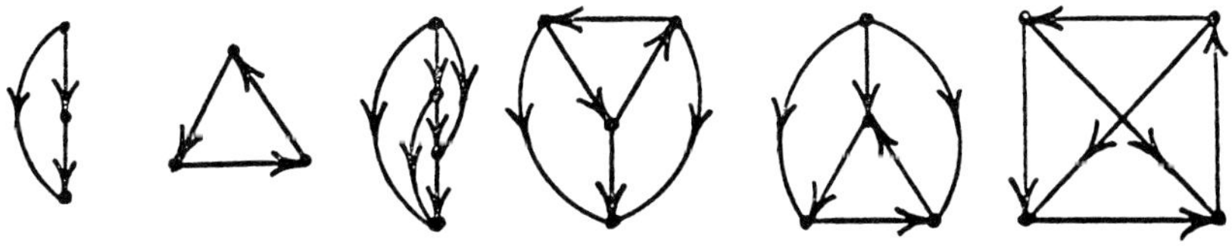

Figure 1.

A path is a sequence of arcs of the type $\overrightarrow{ab}$, $\overrightarrow{bc}$, ..., $\overrightarrow{\ell m}$; we usually assume the nodes a, b, c, ..., ℓ are distinct. If a = m the path is a cycle; an h-cycle is a cycle with h arcs. A tournament T is strong, or strongly connected, if and only if for every ordered pair of nodes p and q of T, there exists a path from p to q in T. It is easy to show (see, for example, [17; p. 5]) that a tournament T is strong if and only if its nodes cannot be partitioned into two non-empty subtournaments A and B such that $A \rightarrow B$; if such subtournaments A and B exist we say T is reducible and write $T = A+B$.

Our object here is to discuss some results that can be formulated in terms of mappings of the arcs or nodes of tournaments; in what follows all such mappings will be one-to-one. We shall frequently use the same symbol to denote the set of nodes of a tournament and the tournament itself.

2. The Automorphism Group of a Tournament.

Let α denote a dominance-preserving permutation of the nodes of a tournament T so that $\alpha(p) \rightarrow \alpha(q)$ if and only if $p \rightarrow q$. The set of all such permutations forms a group, the automorphism group $G(T)$ of T. The following result follows immediately from the definition of a reducible tournament (we denote the direct product of the groups F and H by $F \times H$).

Theorem 2.1. If $T = R + S$, then $G(T) = G(R) \times G(S)$.

Suppose the tournaments R and S have r and s nodes. The composition of R with S is the tournament $R \circ S$ with rs nodes (i, k), where $1 \leq i \leq r$ and $1 \leq k \leq s$, such that $(i, k) \rightarrow (j, \ell)$ if and only if $i \rightarrow j$ in R or $i = j$ and $k \rightarrow \ell$ in S. In other words, $R \circ S$ is obtained by replacing each node i of R by a copy S_i of S and letting $S_i \rightarrow S_j$ in $R \circ S$ if and only if $i \rightarrow j$ in R. The composition of two 3-cycles is illustrated in Figure 2.

Let F and H denote two permutation groups with object sets U and V. The composition (or wreath product) of F with H is the group $F \circ H$ of all permutations α of $U \times V = \{(x, y): x \in U, y \in V\}$ of the type

$$\alpha(x, y) = (f(x), h_x(y)),$$

where f is any element of F and h_x, for each x, is any element of H. If the objects of $U \times V$ are arranged in a

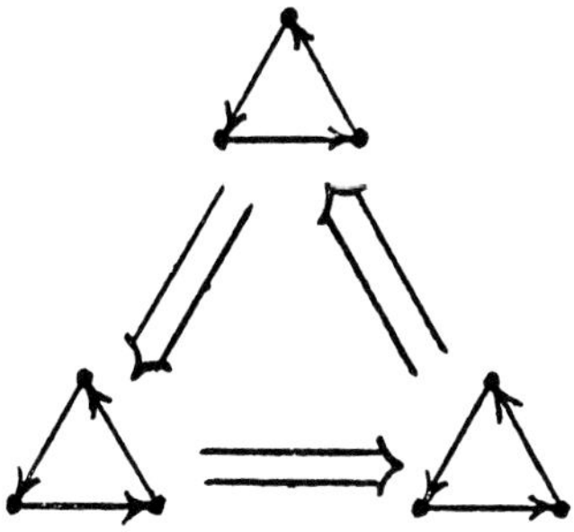

Figure 2

matrix so that the rows and columns correspond to the objects of U and V, then $F \circ H$ is the group of permutations obtained by permuting the objects in each row according to some element of H (not necessarily the same element for every row) and then permuting the rows according to any element of F. The following result was proved by Alspach, Goldberg, and Moon [2].

Theorem 2.2. If R and S are any two tournaments, then

$$G(R \circ S) = G(R) \circ G(S) .$$

Every element of $G(R) \circ G(S)$ is clearly an automorphism of $R \circ S$; the proof that every automorphism of $R \circ S$ is of this type involves an argument that is similar to that used in the proof of Theorem 3.3. (Sabidussi [20] and Hemminger [11] have considered the problem of characterizing those pairs of ordinary graphs, finite and infinite, for which an analogous result holds.)

We now describe the automorphism group of some specific tournaments. In what follows C_m denotes the cyclic group of order m and I denotes the identity group. Let K_n denote the tournament with n nodes whose arcs are defined as follows: if n is odd then $i \to j$ if and only if $0 < j-i \leq (n-1)/2$ where the subtraction is modulo n; if n is even then $i \to j$ if and only if $0 < j-i \leq n/2$ where the subtraction is modulo n+1.

Theorem 2.3. If $n > 2$ is odd then $G(K_n) = C_n$; if n is even then $G(K_n) = I$.

Proof: Suppose n is odd. If $\alpha(i) = i+1$ for $1 \leq i \leq n$ the permutation α^m is certainly an automorphism of K_n for $1 \leq m \leq n$ (the labels are reduced modulo n when necessary). It is not difficult to verify that for each node j the node $j+1$ is the only node k in $\Gamma(j)$ such that the arc $\vec{jk}$ belongs to just one 3-cycle. Thus if β is any automorphism of K_n such that $\beta(i) = j$ then $\beta(i+1) = j+1$. It now follows by induction that the elements of C_n, generated by α, are the only automorphisms of K_n.

Now suppose n is even. The nodes $1, 2, \ldots, n/2$ have score $n/2$ and the nodes $n/2+1, \ldots, n$ have score $n/2-1$. Hence, any automorphism of K_n must permute the first $n/2$ nodes and the last $n/2$ nodes amongst themselves. There are no nontrivial automorphisms with this property because in the subtournament determined by either of these subsets of nodes no two nodes have the same score. This suffices to complete the proof of the theorem.

If $q = p^n$ where n is an odd integer and p is a prime such that $p \equiv 3 \pmod 4$, let R_q denote the tournament whose q nodes are identified with the elements of the Galois field $GF(q)$ and $i \to j$ if and only if $j-i$ is a square in $GF(q)$. Let $S(n, p)$ denote the group of the $nq(q-1)/2$ permutations α of the elements of $GF(q)$ of the type $\alpha(x) = s\beta(x)+b$ where s is a nonzero square of $GF(q)$, β is an automorphism of $GF(q)$, and b is arbitrary in $GF(q)$. The following result was proved by Goldberg [6]; it also follows from a recent result due to W. M. Kantor.

Theorem 2.4. If $q = p^n \equiv 3 \pmod 4$, then

$$G(R_q) = S(n, p) .$$

Let R_∞ denote the tournament with nodes $0, 1, 2, \ldots$ such that $i \to j$ if and only if $i < j$ and $j-i$ is a perfect square or $i > j$ and $i-j$ is not a perfect square.

Theorem 2.5. $G(R_\infty) = I$.

Proof: If p is one of the nodes $0, 1, 2, 3,$ or 4

there is a (unique) node q in $\Gamma(p)$ that is dominated by all the remaining nodes of $\Gamma(p)$; the nodes q corresponding to these nodes p are 1, 2, 0, 1 and 2, respectively. If p is one of the remaining nodes then it is not difficult to verify that each node of $\Gamma(p)$ dominates at least one other node of $\Gamma(p)$. Consequently, any automorphism of R_∞ must permute the nodes 0, 1, 2, 3 and 4 amongst themselves. The subtournament determined by these five nodes has no nontrivial automorphisms so any automorphism of R_∞ leaves these first five nodes fixed. The subtournament determined by the remaining nodes is isomorphic to R_∞; we can repeat this argument, therefore, and it follows that R_∞ has no nontrivial automorphisms.

Theorem 2. 6. A finite group G of order m is abstractly isomorphic to the automorphism group of some tournament if and only if m is odd.

If the group of some tournament has even order it contains at least one self-inverse element α besides the identity element. Hence there exist nodes p and q such that $p \to q$, $\alpha(p) = q$, and $\alpha(q) = p$; this is impossible because it implies that $\alpha(q) \to \alpha(p)$. Thus if G is to be isomorphic to the group of some tournament it is necessary that m be odd. Moon [16] showed this condition is also sufficient by transforming the Cayley color graph of G into a tournament without losing any automorphisms or introducing any new ones. (If G is generated by h of its elements then this construction will yield a tournament with $h(h+1)m/2+m$ nodes. When G is Abelian a much more efficient construction can be based upon Theorems 2. 1 and 2. 3 and the fact that any Abelian group is isomorphic to a direct product of cyclic groups.)

Let $g(T_n)$ denote the order of the group $G(T_n)$ and let $g(n)$ denote the maximum of $g(T_n)$ taken over all tournaments T_n with n nodes. (It is not difficult to show that $g(n)$ is the order of the largest subgroup of odd order of the symmetric group of degree n.) If $C(3)$ denotes a 3-cycle and $C(3^k) = C(3^{k-1}) \circ C(3)$ for $k \geq 2$, then the graph $C(n)$ has order $\sqrt{3}^{\,n-1}$ by Theorem 2. 2 if $n = 3^k$; hence $g(n)$ is at least as large as $\sqrt{3}^{\,n-1}$ if $n = 3^k$.

Theorem 2.7. If $n \geq 1$, then $g(n) \leq \sqrt{3}^{\,n-1}$ with equality holding if and only if $n = 3^k$ for some integer k.

If p is any node in a tournament T_n, let T_d and T_{n-d} denote the subtournaments determined by the nodes of T_n that are and are not similar to p with respect to $G(T_n)$; it follows that

$$g(T_n) \leq g(T_d) \cdot g(T_{n-d}) < g(d) \cdot g(n-d)$$

if $1 \leq d \leq n-1$. If p is similar to every node of T_n (this can happen only if n is odd and all nodes of T_n have the same score), then it follows from an elementary result in group theory that

$$g(T_n) = ng(H),$$

when $g(H)$ denotes the order of the subgroup H of automorphisms α of $G(T_n)$ such that $\alpha(p) = p$. Since no element of H can transform one of the $(n-1)/2$ nodes dominating p into one of the $(n-1)/2$ nodes p dominates it follows that

$$g(T_n) \leq n\{g((n-1)/2)\}^2$$

in this case.

Goldberg and Moon [7] used these two upper bounds for $g(T_n)$ to show by induction that $g(n) \leq (2.5)^n/2n$ for $n \geq 4$. They conjectured that the inequality in Theorem 2.7 holds and showed that if

$$c = \lim_{n \to \infty} g(n)^{1/n}$$

then c exists and $\sqrt{3} \leq c \leq 2.5$. (The proof of this result makes use of the inequality $g(ab) \geq g(a)[g(b)]^a$, for any positive integers a and b; this inequality follows from Feit-Theorem 2.2.) Dixon [4] subsequently proved Theorem 2.7 by a group-theoretic argument that made use of the Thompson theorem. Alspach [1] later gave a purely combinatorial proof by induction that involved a refinement of the bound stated above for $g(T_n)$ when all nodes of T_n are similar.

3. When is the Composition of Two Tournaments Commutative?

The composition of two tournaments is not commutative in general. For example, let R and S denote strong tournaments with three and four nodes; there are nodes in $S \circ R$ with score seven but there are no such nodes in $R \circ S$. Goldberg and Moon [8] have investigated some algebraic properties of the composition operation in order to characterize those pairs of nontrivial tournaments R and S such that $R \circ S = S \circ R$. We shall outline a proof of some of their results in the special case when the tournaments are strong; most of the proofs in the general case are considerably more complicated.

Theorem 3.1. If A, B, and C are any three tournaments, then $A \circ (B \circ C) = (A \circ B) \circ C$.

Theorem 3.2. If $|A| > 1$ then $A \circ B$ is strong if and only if A is strong.

These two results are straightforward consequences of the relevant definitions.

Theorem 3.3. Let α denote a one-to-one dominance preserving mapping of the nodes of $A \circ B$ onto the nodes of $Y \circ Z$. If Z is strong and $|B| \leq |Z|$, then for each copy B_i of B in $A \circ B$ there is a copy Z_j of Z in $Y \circ Z$ such that $\alpha(B_i) \subseteq Z_j$.

Proof. Suppose, on the contrary, that there exists a copy B_i such that some but not all the nodes of $\alpha(B_i)$ belong to Z_j. If X denotes the set of nodes of B_i that are mapped into Z_j by α, then both $B_i - X$ and $Z_j - \alpha(X)$ are nonempty.

If any node z of $Z_j - \alpha(X)$ both dominates and is dominated by nodes of $\alpha(X)$, then $\alpha^{-1}(z)$ has the same property with respect to the nodes of X. The only nodes in $A \circ B$ with this property belong to B_i; if $\alpha^{-1}(z)$ belonged to B_i then z would belong to $\alpha(X)$ contrary to the definition of z. Hence, for each node z of $Z_j - \alpha(X)$ either $z \rightarrow \alpha(X)$ or $\alpha(X) \rightarrow z$. Since Z_j is strong, by assumption, there must exist two nodes p and q of $Z_j - \alpha(X)$ such that

$p \rightarrow \alpha(X)$ and $\alpha(X) \rightarrow q$.

The nodes $\alpha^{-1}(p)$ and $\alpha^{-1}(q)$ do not belong to B_i in $A \circ B$; consequently, $\alpha^{-1}(p) \rightarrow B_i$ and $B_i \rightarrow \alpha^{-1}(q)$. In particular, if u is any node of $B_i - X$ then $\alpha^{-1}(p) \rightarrow u$ and $u \rightarrow \alpha^{-1}(q)$. The node $\alpha(u)$ does not belong to Z_j in $Y \circ Z$ so either $\alpha(u)$ dominates both p and q or both p and q dominate $\alpha(u)$ in $Y \circ Z$. This violates the hypothesis that α is dominance preserving. Thus $B_i - X$ must be empty and the theorem is proved. (The same conclusion can be reached in essentially the same way if we assume B instead of Z is strong.)

Corollary 3.3.1. If the tournaments A, Y, and Z are strong, then

$$A \circ Y = A \circ Z \Longleftrightarrow Y = Z$$

and

$$Y \circ A = Z \circ A \Longleftrightarrow Y = Z .$$

Theorem 3.4. If $A \circ B = Y \circ Z$ where Z is strong and $|B| \leq |Z|$, then there exists a tournament W such that $Z = W \circ B$.

Proof. If α is defined as in Theorem 3.3 then it follows that there exist integers $i(1), \ldots, i(t)$ such that $\alpha(\bigcup B_{i(h)}) = Z_1$ where $h = 1, \ldots, t$. If W denotes the subtournament of A determined by the nodes $i(1), \ldots, i(t)$ then $Z = Z_1 = W \circ B$ and the theorem is proved. (If $|Z| = |B|$, then W is the trivial tournament consisting of a single node.)

In the present context we say a tournament T is prime if $|T| \geq 3$, T is strong, and there do not exist nontrivial strong tournaments A and B such that $T = A \circ B$.

Theorem 3.5. If $A \circ B = Y \circ Z$ where B and Z are prime tournaments, then $A = Y$ and $B = Z$.

Proof. We may suppose $|B| \leq |Z|$. If $B \neq Z$, there exists a nontrivial tournament W such that $Z = W \circ B$ by Theorem 3.4. Since Z is strong W must also be strong by Theorem 3.2. This contradicts the hypothesis that Z is prime; therefore $B = Z$ and $A = Y$ by Corollary 3.3.1.

This result implies the following unique factorization theorem. (Sabidussi [19], Lovász [15], and Imrich [12] have obtained analogous results for ordinary graphs with respect to various operations.)

Theorem 3.6. If T is a nontrivial strong tournament then there exist prime tournaments $R_1, \ldots, R_m$ such that $T = R_1 \circ \ldots \circ R_m$; furthermore, if $T = S_1 \circ \ldots \circ S_n$ where the tournaments $S_1, \ldots, S_n$ are all prime, then $m = n$ and $R_i = S_i$ for $i = 1, \ldots, m$.

Let $T^1 = T$ and $T^n = T \circ T^{n-1}$ for $n = 2, 3, \ldots$.

Theorem 3.7. If R and S are nontrivial strong tournaments, then $R \circ S = S \circ R$ if and only if there exists a tournament T and integers m and n such that $R = T^m$ and $S = T^n$.

Proof. We may suppose $|S| \leq |R|$ then $S = R$ by Theorem 3.3. If $|S| < |R|$ then $R = W \circ S$ for some nontrivial strong tournament W. In this case $W \circ S \circ S = S \circ W \circ S$ or $W \circ S = S \circ W$ by Theorem 3.1 and Corollary 3.3.1. The result now follows by induction on $\max(|S|, |R|)$.

A tournament T_n is transitive if its n nodes can be labelled in such a way that $i \rightarrow j$ if and only if $i > j$. If we drop the assumption that R and S are strong in Theorem 3.7, then the conclusion must be amended to include the possibility that R and S are both transitive. (We remark that Imrich [13] has characterized those pairs of ordinary graphs whose composition is commutative.)

4. Arc Mappings, Isomorphisms, and Anti-Isomorphisms.

Let θ denote a one-to-one correspondence between the edges of two (finite) graphs G and H. Whitney [21] showed that if θ maps adjacent edges of G onto adjacent edges of H and conversely, then θ is induced by an isomorphism between G and H if these graphs are connected and have at least five nodes (two edges are adjacent if they have a node in common). Whitney also showed (see also Ore [18; pp. 245-249]) that if G and H are 3-vertex connected and θ maps cycles of G onto cycles of H and conversely, then θ is also induced by an isomorphism between G and H. Jung [14] extended the first result to infinite graphs and Halin and Jung [10] proved a more general result that contains both of these results as special cases. Goldberg and Moon [9] recently proved a result of this type for tournaments.

Let ϕ denote a (one-to-one) mapping of the arcs of a tournament R onto the arcs of a tournament S (in what follows we denote the arc joining nodes i and j by ij or ji when we do not want to specify its orientation). We say R and S are <u>h-cycle isomorphic</u> (with respect to ϕ) if any h arcs in one tournament form a cycle if and only if the corresponding arcs in the other tournament form a cycle (but not necessarily in the same order).

<u>Theorem 4.1.</u> If two strong tournaments R and S are 3-cycle and 4-cycle isomorphic then they are either isomorphic or anti-isomorphic.

The conclusion does not necessarily hold if R and S are not strong. For example, if T_3 and T_4 denote strong tournaments with three and four nodes, let $R = T_3 + T_4 + T_4$ and $S = T_4 + T_3 + T_4$; it is easy to define a mapping ϕ from the arcs of R onto the arcs of S such that R and S are 3-cycle and 4-cycle isomorphic with respect to ϕ yet they are neither isomorphic nor anti-isomorphic.

Neither does the conclusion follow if R and S are not 4-cycle isomorphic. Let R denote the tournament formed from three subtournaments A, B, and C by letting $A \rightarrow B$, $B \rightarrow C$, and $C \rightarrow A$; let S differ from R in that $A \rightarrow C$, $C \rightarrow B$, and $B \rightarrow A$. The tournaments R and S are strong and they are 3-cycle isomorphic with respect to the arc mapping ϕ induced by an isomorphism between the copies of A, B, and C in the two tournaments. For most choices of A, B, and C, however, R and S are not 4-cycle isomorphic and they are neither isomorphic nor anti-isomorphic (for a specific example, let A and B denote tournaments with one and two nodes and let C denote a tournament of four nodes consisting of a 3-cycle each node of which is dominated by the fourth node). The proof of Theorem 4.1 is based on another result that is formulated in terms of a stronger appearing hypothesis.

We say a (nontrivial) subtournament M of R has <u>property $P(\phi)$</u> if ϕ induces an isomorphism or anti-isomorphism between M and some subtournament N of S, that is, if there exists a mapping $\alpha = \alpha_M$ of the nodes of M onto the nodes of N such that

$$\phi(ij) = \alpha(i)\alpha(j) \quad \text{for all } i, j \in M, \text{ and either}$$

$$i \rightarrow j \Longleftrightarrow \alpha(i) \rightarrow \alpha(j) \quad \text{for all } i, j \in M \text{ or}$$

$$i \rightarrow j \Longleftrightarrow \alpha(j) \rightarrow \alpha(i) \quad \text{for all } i, j \in M .$$

If $h \geq 4$ and all (nontrivial) strong subtournaments of R and S with at most h nodes have properties $P(\phi)$ and $P(\phi^{-1})$, respectively, then we say R and S are <u>h-equivalent</u> with respect to ϕ.

<u>Theorem 4.2.</u> If $h \geq 4$ and R and S are h-equivalent with respect to ϕ, then there exists a mapping ψ of the arcs of R onto the arcs of S such that R and S are $(h+1)$-equivalent with respect to ψ; furthermore, ψ and ϕ agree for all arcs contained in strong subtournaments with at most h nodes.

It is not difficult to show that if two tournaments R and S are 3-cycle and 4-cycle isomorphic they are 4-equivalent. If $|R| = |S| = n \geq 4$, then it follows by induction from Theorem 4.2 that R and S are n-equivalent with respect to some mapping ψ; hence, if R is strong then R itself has property $P(\psi)$ and this implies the conclusion of Theorem 4.1.

The proof of Theorem 4.2 given in [9] is too long to be repeated here but it applies even when R and S have an enumerable number of nodes; we now show that Theorem 4.1 follows from Theorem 4.2 in this case also.

Suppose the nodes of the strong infinite tournaments R and S are labelled $1, 2, \ldots$. Let $T_1, T_2, \ldots$ denote an infinite sequence of nontrivial, finite, strong subtournaments of R such that

(a) node i belongs to tournament T_i, and

(b) $T_1 \subseteq T_2 \subseteq T_3 \subseteq \cdots$.

Such a sequence can be defined inductively as follows: let T_1 denote any nontrivial, finite, strong subtournament that contains node 1; if k is the node with the smallest label not contained in T_i, let T_{i+1} denote any minimal finite strong subtournament that contains the nodes of T_i and k. The fact that there is a finite path from p and q for any nodes p and q of R implies such subtournaments exist.

If R and S are 3-cycle and 4-cycle isomorphic then they are 4-equivalent, as before, and every arc of R belongs to some finite strong subtournament of R. If we apply Theorem 4.2 when $h = 4, 5, \ldots,$ we conclude that there exists a mapping ψ of the arcs of R onto the arcs of S such that every finite strong subtournament of R has property $P(\psi)$. It is not difficult to show (see [9]) that if two subtournaments K and L of R both have property $P(\psi)$ and $K \subseteq L$, then the node mappings α_K and α_L of K and L induced by ψ agree on the nodes of K; furthermore α_K and α_L both define isomorphisms or they both define anti-isomorphisms between K and L and some subtournaments of S.

If x is any node of R let $\alpha(X) = \alpha_T(X)$ where α_T is the node mapping induced by ψ on the first subtournament T of the sequence $T_1, T_2, \ldots$ that contains x. The subtournaments $T_1, T_2, \ldots$ all have property $P(\psi)$ and it follows from the observation at the end of the last paragraph that this node mapping α defines an isomorphism or anti-isomorphism between R and some subtournament N of S. It must be that the subtournament N is actually S itself since α is induced by ψ and ψ is a one-to-one mapping of the arcs of R onto the arcs of S. This suffices to show that Theorem 4.1 follows from Theorem 4.2 when R and S have an enumerable number of nodes. (Notice that if the strong tournaments R and S are 3-cycle and 4-cycle isomorphic with respect to the arc mapping ϕ we cannot conclude that ϕ induces an isomorphism or anti-isomorphism between R and S; we can only conclude that there exists an arc mapping ψ that induces an isomorphism or anti-isomorphism where ψ agrees with ϕ on all arcs that belong to 3-cycles or 4-cycles.)

5. Minimal Regular Tournaments Containing a Given Tournament.

An ordinary graph is r-regular if each node is incident with r edges; an oriented or directed graph is r-regular if there are r arcs leading towards and r arcs leading away from each node. Let G denote a graph in which the maximum number of edges incident with any node is r; Erdös and Kelly [5] determined the least integer m for which there exists an r-regular graph H with m nodes such that some subset of nodes of H determines a subgraph isomorphic to G. Beineke and Pippert [3] solved the corresponding problem for oriented and directed graphs. In both these papers the size of the smallest regular graph containing the given graph is determined by finding the smallest number that satisfies three inequalities. When the graph is a tournament two of these inequalities are unnecessary; we prove a result equivalent to Beineke and Pippert's for this particular case.

Let s_i and ℓ_i denote the number of arcs oriented away from and towards the i-th node of a tournament T_k with k nodes so that $s_i + \ell_i = k-1$ for each i. We may suppose the nodes are labelled so that $s_1 \leq s_2 \leq \cdots \leq s_k$ and $\ell_1 \geq \ell_2 \geq \cdots \geq \ell_k$. Let $n(T_k)$ denote the least integer n for which there exists a regular tournament T_n that contains a subtournament isomorphic to T_k, assuming that such a regular tournament exists.

Theorem 5.1. If T_k is any tournament, then

$$n(T_k) = 2 \max \{s_k, \ell_1\} + 1.$$

Proof. We first suppose $s_k \geq \ell_1$. Each node in a regular tournament T_n has score $(n-1)/2$; if T_n contains (a subtournament isomorphic to) T_k then $(n-1)/2 \geq s_k$ or $n \geq 2s_k+1$. We shall now show that there exists a regular tournament T_n with $n = 2s_k+1$ nodes that contains T_k.

If $m = 2s_k+1-k$ is odd let K_m denote the regular tournament with m nodes defined on page 135 in which each node has score $(m-1)/2$. The tournament T_n will be constructed by orienting km arcs that collectively join each node of T_k with each node of K_m in such a way that the resulting tournament with $k+m=n$ nodes is regular.

If $r_i = s_k - s_i$ for $1 \leq i \leq k$, let the first node of T_k dominate the first r_1 nodes of K_m, let the second node dominate the next r_2 nodes, and so on (whenever we reach the last node of K_m we return to the first node and continue). This is possible without having more than one arc oriented from any node in T_k to any node in K_m since

$$r_i = s_k - s_i \leq s_k - s_1 = s_k + \ell_1 + 1 - k \leq 2s_k + 1 - k = m$$

for each i.

At each stage in this process the numbers of arcs oriented from nodes of T_k to any two nodes of K_m differ by at most one. When the process is completed suppose α nodes of K_m have t arcs oriented towards them and the remaining $m-\alpha$ nodes have $t+1$ arcs oriented towards them (from nodes of T_k). The total number of arcs oriented from nodes of T_k to nodes of K_m is

$$\sum_{i=1}^{k} r_i = \sum_{i=1}^{k} (s_k - s_i) = ks_k - \binom{k}{2} = \frac{1}{2} km ;$$

therefore,

$$\frac{1}{2} km = \alpha t + (m-\alpha)(t+1) = mt + (m-\alpha)$$

and either $\alpha = 0$ or $\alpha = m$ since k is even. This implies that there are $k/2$ arcs oriented from nodes of T_k to each node of K_m. Let each node of K_m dominate the $k/2$ nodes of T_k that do not already dominate it.

In the tournament T_n thus defined the i-th node of T_k has score $s_i + r_i = s_k$ and each node of K_m has score $k/2 + (m-1)/2 = s_k$. Hence, T_n is a regular tournament with $n = k+m = 2s_k+1$ nodes that contains T_k. This completes the proof of the theorem when $s_k \geq \ell_1$ and m is odd; only slight changes are necessary in the preceding argument to take care of the case when m is even.

If $\ell_1 > s_k$ let T_n' denote the tournament obtained by applying the preceding construction to the tournament obtained from T_k by reversing the orientation of all its arcs. If T_n denotes the tournament obtained from T_n' by reversing all its arcs then T_n is a regular tournament with $n = 2\ell_1+1$ nodes that contains T_k .

REFERENCES

1. B. Alspach, A combinatorial proof of a conjecture of Goldberg and Moon, Can. Math. Bull. 11 (1968) 655-661.

2. B. Alspach, M. Goldberg, and J. W. Moon, The group of the composition of two tournaments, Math. Mag. 41 (1968) 77-80.

3. L. W. Beineke and R. E. Pippert, Minimal regular extensions of oriented graphs, Amer. Math. Monthly 76 (1969) 145-151.

4. J. D. Dixon, The maximum order of the group of a tournament, Can. Math. Bull. 10 (1967) 503-505.

5. P. Erdős and P. Kelly, The minimal regular graph containing a given graph, Amer. Math. Monthly 70 (1963) 1074-1075.

6. M. Goldberg, The group of the quadratic residue tournament, Can. Math. Bull. (to appear).

7. M. Goldberg and J. W. Moon, On the maximum order of the group of a tournament, Can. Math. Bull. 9 (1966) 563-569.

8. M. Goldberg and J. W. Moon, On the composition of two tournaments, Duke Math. J. (to appear).

9. M. Goldberg and J. W. Moon, Arc mappings and tournament isomorphisms, (unpublished manuscript).

10. R. Halin and H. A. Jung, Note on isomorphisms of graphs, J. Lond. Math. Soc. 42 (1967) 254-256.

11. R. L. Hemminger, The group of an X-join of graphs, J. Combinatorial Theory 5 (1968) 408-418.

12. W. Imrich, Kartesisches Produkt von Mengensystemen und Graphen, Stud. Sci. Math. Hung. 2(1967) 285-290.

13. W. Imrich, Über das lexikographische Produkt von Graphen, Arch. Math. 20 (1969) 228-234.

14. H. A. Jung, Zu einem Isomorphiesatz von H. Whitney für Graphen, Math. Ann. 164 (1966) 270-271.

15. L. Lovász, Über die starke Multiplikation von Geordenten Graphen, Acta Math. Acad. Sci. Hung. 18 (1967) 235-241.

16. J. W. Moon, Tournaments with a given automorphism group, Can. J. Math. 16 (1964) 485-489.

17. J. W. Moon, Topics on Tournaments, New York, Holt, Rinehart and Winston, 1968.

18. O. Ore, Theory of Graphs, Providence, A.M.S. Colloq. Pub., 38, 1962.

19. G. Sabidussi, Graph multiplication, Math. Zeit. 72 (1960) 446-457.

20. G. Sabidussi, The lexicographic product of graphs, Duke Math. J. 28 (1961) 573-578.

21. H. Whitney, Congruent graphs and the connectivity of graphs, Amer. J. Math. 54(1932) 150-168.

The University of Alberta
Edmonton, Alberta
Canada

On Some Connections between Graph Theory and Experimental Designs and Some Recent Existence Results

D. K. RAY CHAUDHURI

1. Introduction.

To compare the effects of a number of treatments, the experimenter chooses a set of experimental units, applies the treatments to the units according to a plan or design, observes the responses of the experimental units after some time and makes his conclusion on the basis of the observed responses. The observed responses are also affected by extraneous factors which vary over the experimental units. Since the observed responses vary even if the experiment is repeated under similar conditions, the experimenter postulates a probabilistic model. It is assumed that the response is really a random variable whose distribution depends on the treatments applied as well as a number of extraneous factors. A design of the experiment consists of the choice of the given number of experimental units and the assignment of treatments to the units. The amount of information derivable from the experiment can be increased by giving proper consideration to the extraneous factors. Optimum experimental designs often require complicated arrangements of the experimental units according to various factors and such arrangements are based on combinatorial configurations or combinatorial designs. Both graphs and combinatorial designs are special cases of incidence structures. In a way graphs are simpler incidence structures. Often problems about combinatorial designs can be translated into problems about graphs. In recent years several authors applied graph theory to prove existence theorems as well as non-existence theorems about combinatorial designs. In this paper we present a review of some of these results without proofs. We also try to formalize and clarify some of the basic concepts of experimental designs.

2. Description of an experimental design.

In an experiment there will be a set of treatments $\mathfrak{T} = (T_1, T_2, \cdots, T_v)$ and a set of experimental units $\mathfrak{E} = (E_1, E_2, \cdots, E_\nu)$. The experimental units are not all identical. A set of extraneous variables $X = (x_1, x_2, \cdots, x_p)$ take different values on the set $\mathfrak{E}$. $H \subseteq R_p$, (the p dimensional Euclidean space) is the set of possible values of X. For each $T \in \mathfrak{T}$ and $a \in H$, $\mathfrak{F}_{T, a}$ is a specified class of probability distribution functions. The experiment is based on a mapping $D: \mathfrak{E} \rightarrow \mathfrak{T}$. The mapping D is usually called the experimental design. The treatment $D(E)$ is applied to the experimental unit E and the response $y(D, E)$ is observed. It is assumed that $y(D, E)$ is a random variable whose distribution function belongs to the class $\mathfrak{F}_{T, a}$ where $D(E) = T$ and $X(E) = a$. The purpose of the experiment is to make a decision about the set of treatments, i. e., about the classes of distribution functions $\{\mathfrak{F}_{T, a}: T \in \mathfrak{T}, a \in H\}$. The "decision function" itself will be a function of the random variables $Y(D, E)$, $E \in \mathfrak{E}$ and hence the "goodness" of the decision function depends on the experimental design D. To give an example, in a nutritional experiment for selecting the best diet for pigs the treatments are four different diets. Treatments are applied to pigs taken from 5 different litters. The extraneous variable X corresponds to the litters and take 5 distinct values. Response is gain in the weight of a pig. y_{ij}, the gain in the weight of the jth pig from the ith litter, is assumed to be a random variable with mean $\beta_i + \tau_j$ where β_i represents the effect of the ith litter and τ_j represents the effect of the jth diet, $i = 1, 2, 3, 4, 5$ and $j = 1, 2, 3, 4$. It is also assumed that the variance of y_{ij} is σ^2 and that y_{ij} and $y_{i'j'}$ are uncorrelated for $(i, j) \neq (i', j')$, $i, i' = 1, \cdots, 5$ and $j, j' = 1, 2, 3, 4$.

3. Incidence structures, graphs and balanced incomplete block designs.

An incidence structure is a triple (P, L, I) where P is a set of objects called points or treatments or vertices, L is another set of objects called lines or blocks or edges and I is a subset of $P \times L$. P and L are assumed to be disjoint. If for $p \in P$, $\ell \in L$, $(p, \ell) \in I$, the points p and ℓ are said to be incident with each other. A graph is an incidence structure in which every edge is incident with one vertex or two vertices. An edge which is incident with one vertex is called a loop. If there are two or more edges ℓ_1 and ℓ_2 incident with the same set of two vertices p_1 and p_2, then we say there are multiple edges between p_1 and p_2. Two vertices p_1 and p_2, not necessarily distinct, are said to be adjacent if and only if there is an edge incident with p_1 and p_2. A graph G without loops and multiple edges can be described as $G = (P, E)$ where $E \subseteq P^{(2)}$, $P^{(2)}$ being the set of unordered pairs of elements of P. The unordered pair $(p_1, p_2) \in E$ if and only if there is an edge ℓ incident with p_1 and p_2.

Let v and λ be positive integers and K be a set of positive integers. Then a (v, K, λ)-pairwise balanced design (PBD) is an incidence structure (P, L, I) where

(i) $|P| = v$,

(ii) For every $\ell \in L$, ℓ is incident with k points where $k \in K$,

and

(iii) For $p_1, p_2 \in P$, $p_1 \neq p_2$, there exist exactly λ distinct lines ℓ_i, $i = 1, 2, \cdots, \lambda$ such that ℓ_i is incident with both p_1 and p_2.

If K consists of a single integer k, then the PBD is called a (v, k, λ)-balanced incomplete block design (BIBD). Usually the line (or block) ℓ is identified with the set $A_\ell = \{p \mid (p, \ell) \in I\}$ of points. Then the BIBD is considered to be the pair $(P, \mathfrak{A})$ where $\mathfrak{A}$ is the family of blocks $(A_\ell : \ell \in L)$. The following is a BIBD with $v = 7$, $k = 3$, $\lambda = 1$, $P = \{1, 2, 3, 4, 5, 6, 7\}$:

$A_1 = (1, 2, 3)$, $A_2 = (1, 4, 5)$, $A_3 = (1, 6, 7)$, $A_4 = (2, 4, 6)$, $A_5 = (2, 5, 7)$, $A_6 = (3, 4, 7)$ and $A_7 = (3, 5, 6)$.

In a (v, k, λ)-BIBD, the number of blocks must be equal to $b = \frac{\lambda v(v-1)}{k(k-1)}$ and every treatment (or point occurs in $r = \frac{\lambda(v-1)}{k-1}$ blocks. In a (v, k, λ)-balanced incomplete block design (BIBD), there will be a set of bk experimental units, one extraneous variable (or factor) X which takes b distinct values $a_1, a_2, \cdots a_b$, and a set $P = \{p_1, p_2, \cdots, p_v\}$ of v treatments. The "group" $B_i = \{E \mid X(E) = a_i\}$ contains k experimental units, $i = 1, 2, \cdots, b$. $\mathfrak{E} = \bigcup_{i=1}^{b} B_i$. The b "groups" of experimental units are first assigned to the b blocks of the (v, k, λ)-BIBD and finally the k units of the group assigned to the k treatments of the particular block. To be more formal, let $I_b = \{1, 2, \ldots, b\}$. Let σ be a bijection (one-to-one onto mapping) $I_b \to I_b$ and τ_i be a bijection $B_i \to A_{\sigma(i)}$. For every choice of σ and $(\tau_i, i = 1, 2, \ldots, b)$, we have a BIB experimental design D defined by $D(E) = \tau_i(E)$ when $E \in B_i$. In the literature on experimental design, frequently no distinction is made between the combinatorial design and the experimental design. It seems that for the sake of conceptual clarity, such a distinction is very useful. The (7, 3, 1)-BIBD could be used for an experiment to select the "best" among 7 different kinds of ice creams (7 treatments), there being 7 human tasters each of whom tries three different ice creams consecutively and gives each of them a score. The score given to a particular ice cream depends on the quality of the ice cream as well as the taster. In this experiment the 7 tasters represent the 7 "levels" (or values) of the extraneous factor. However it seems reasonable to assume that the difference of the scores given to a pair of ice creams will not be affected by the taster. Under this assumption the experiment provides more information if we can ensure that every pair of ice creams is tasted by a common taster an equal number of times. This property is satisfied by the BIB experimental design. Kiefer [8] proved that within a certain class of experimental designs, the BIB experimental design is optimum with respect to several optimality criteria. Let $\mathfrak{E} = \{E_1, E_2, \ldots, E_\nu\}$, $\nu = vr$ be a set of ν experimental units, let x be an extraneous factor that takes b values (or levels) a_i, $i = 1, 2, \ldots, b$ and $P = \{p_1, p_2, \ldots, p_v\}$ be a set of v treatments. Let $\beta = \beta_1, \beta_2, \ldots, \beta_b)$ and $\tau = (\tau_1, \tau_2, \ldots, \tau_v)$ be respectively b-dimensional and v-dimensional real vectors and let σ^2 be a positive real number. For any integer n, $\mathbb{R}^n$ will denote the n-dimensional real space and $\mathbb{R}^+$ will denote the set of positive real numbers. For $\beta \in \mathbb{R}^b$, $\tau \in \mathbb{R}^v$ and $\sigma^2 \in \mathbb{R}^+$, $Y_{\beta, \tau, \sigma^2}$ will denote the class of all ν dimensional random variables $y = (y_{ij}, i = 1, 2, \ldots, b, j = 1, 2, \ldots, v)$ such that

(i) the expectation of y_{ij} is $\beta_i + \tau_j$,

(ii) the variance of y_{ij} is σ^2,

(iii) and the covariance of y_{ij} and $Y_{i'j'}$ is 0 whenever $(i, j) \neq (i', j')$, $i, i' = 1, 2, \ldots, b$, $j, j' = 1, 2, \ldots, v$.

A design D is a mapping from $\mathfrak{E}$ to P. When the treatment $D(E)$ is applied, the response $y(D, E)$ is a random variable with the same distribution as y_{ij} where $D(E) = p_j$ and $X(E) = a_i$, $i = 1, 2, \ldots, b$ and $j = 1, 2, \ldots, v$.

The "parameter" β_i denotes the effect of the ith level of the extraneous factor x and τ_j denotes the effect of the jth treatment p_j. Let $\pi = \sum_{j=1}^{v} a_j \tau_j$, $a_j \in \mathbb{R}$. The "parametric function" π is said to be estimable with respect to a given design D iff there exist real numbers $C(E)$, $E \in \mathfrak{E}$ such that for every $\beta \in \mathbb{R}^b$, $\tau \in \mathbb{R}^v$, $\sigma^2 \in \mathbb{R}^+$ and $y \in Y_{\beta, \tau, \sigma^2}$ the expectation of $\sum_{E \in \zeta} c(E) y(D, E)$ is $\sum_{j=1}^{v} a_j \tau_j$.

The parametric function π is said to be a treatment contrast if and only if $\sum_{j=1}^{v} a_j = 0$. Let $\mathfrak{D}$ be the set of all experimental designs D for which all treatment contrasts are estimable. If a (v, k, λ)-BIB exists, then any BIB experimental design $D_0 \in \mathfrak{D}$ and possesses several optimal properties. We mention only two of these properties [Kiefer, (8)]. Let $\pi_1, \pi_2, \ldots \pi_{v-1}$ be a set of $(v-1)$ linearly independent contrasts and $\hat{\pi}_1, \hat{\pi}_2, \cdots, \hat{\pi}_{v-1}$ be the corresponding minimum variance unbiased estimators. The variance-covariance matrix of these estimators is denoted by $\sigma^2 V_D$. The reciprocal of the determinant $|V_D|$ is a measure of goodness of the design D. Similarly the reciprocal of $\lambda(V_D)$, the maximum eigenvalue of V_D is another measure of goodness of the design. Kiefer showed that for any BIB experimental design D_0,

$$|V_{D_0}| = \min_{D \in \mathcal{D}} |V_D| ,$$

$$\lambda(V_{D_0}) = \min_{D \in \mathcal{D}} \lambda(V_D) .$$

4. The Youden square design and the Kőnig-Egervary therorm.

Historically the first connection between graph theory and experimental designs was through Youden Square (YS) designs. A(v, k, λ)-BIBD is said to be symmetric if and only if $b = v$, where b denotes the number of blocks and is equal to $\frac{\lambda v(v-1)}{k(k-1)}$. A$(v, k, \lambda)$-YS design is an arrangement of v treatments into v rows and k columns such that each treatment occurs in each column exactly once and each pair of distinct treatments occur together in λ distinct rows. The following is a $(7, 3, 1)$-YS design:

1	2	3
2	4	6
3	6	5
4	5	1
5	7	2
6	1	7
7	3	4 .

Clearly the rows of a (v, k, λ)-YS design consititute the blocks of a (v, k, λ)-symmetric BIBD. Conversely the blocks of a symmetric (v, k, λ)-BIBD can be ordered to produce the rows of a (v, k, λ)-YS design. This fact was first proved by Smith and Hartley [12]. However there is an immediate proof by the König-Egervary theorem. Let N be the block-treatment incidence $(v \times v)$ matrix of the BIBD, i. e., $N = ((n_{ij}))$ where $n_{ij} = 1(0)$ if and only if the ith block contains (does not contain) the jth treatment, $i, j = 1, 2, \ldots, v$. It follows easily that each row sum and column sum of the incidence matrix is k. From the König-Egervary theorem it can be established that $N = P_1 + P_2 + \ldots + P_k$ where the P_i's are permutation matrices [see Ryser [11], p. 58]. To construct the YS design on the set of objects $I_v = \{1, 2, \ldots, v\}$, define $y_{i\ell} = j$ when the permutation matrix P_ℓ has a 1 in the ith row and jth column, $i = 1, 2, \ldots, v$; $\ell = 1, 2, \ldots, k$ and note that $Y = ((y_{i\ell}))$ is a YS design. YS designs are useful for constructing efficient experimental designs when the experimental units are subject to the influence of two extraneous factors. There will be vk experimental units $E_1, E_2, \ldots, E_\nu$ where $\nu = vk$ and the two extraneous factors are X_1 and X_2, X_1 taking v values (or levels) $a_1, a_2, \ldots, a_v$ and X_2 taking k values $c_1, c_2, \ldots, c_k$. There is exactly one unit E_{ij} for which $X_1(E) = a_i$ and $X_2(E) = c_j$ $i = 1, 2, \ldots, v$ and $j = 1, 2, \ldots, k$. In the YS experimental design, the rows and columns respectively are assigned to the levels of the first and second extraneous factor. More formally, let $\tau : I_v \to I_v$ and $\sigma: I_k \to I_k$ be bijections. Let $Y = ((y_{i\ell}))$, $i = 1, 2, \ldots, v$ and $\ell = 1, 2, \ldots, k$ be a YS design. Then a YS-experimental design is a mapping $D: \mathfrak{E} \to I_v$ defined by $D(E) = y_{\tau(i), \sigma(\ell)}$ where $X_1(E) = a_i$ and $X_2(E) = c_\ell$, $i = 1, 2, \ldots, v$ $\ell = 1, 2, \ldots, k$. Here I_v is the set of treatments. The (7, 3, 1)-YS design could be used in the ice cream selection experiment. The scores given to the ice creams will probably also be affected by the order in which the particular taster tastes the ice creams. Hence it is reasonable to postulate that there are two extraneous variables, X_1 corresponding to the taster and X_2 corresponding to the order of tasting X_1 takes 7 values and X_2 takes 3 values. Kiefer proved that the YS experimental design enjoy optimally properties similar to those of the BIB experimental design in a suitably defined class of experimental designs.

5. Graph theoretic characterization of symmetric BIBD's and some existence and nonexistence results.

One of the important problems concerning BIBD's is to determine all triples (v, k, λ) for which a (v, k, λ)-BIBD exists. This is an extremely difficult problem. A special case of the problem is the problem of determining all integers n which are orders of a finite projective plane. There is an extensive literature on the subject of finite projective planes (see Hall [4], chapter 12). However we still do not know if there is an integer n which is not a prime power and still is the order of a finite projective plane. All prime powers are known to be orders of finite projective planes. Existence of an $(n^2 + n+1, n+1, 1)$-BIBD implies the existence of a finite projective plane of order n. It is possible to give several graph theoretical characterization of (v, k, λ)-symmetric BIBD. One non-trivial characterization was proved by Hoffman and Ray-Chaudhuri [6]. Let $\Pi = (P, L, I)$ be a symmetric BIBD as defined in section 2. Define $G(\Pi)$ to be a graph without loops and multiple edges with vertex set $V = \{(p, \ell) : (p, \ell) \in I\}$ and two vertices (p, ℓ) and (p', ℓ') being adjacent if and only if $p = p'$ or $\ell = \ell'$. The adjacency matrix of a graph on v vertices is the symmetric matrix $\underset{(v\times v)}{A} = ((a_{ij}))$ where $a_{ij} = 1(0)$ according as the ith vertex are (not) adjacent. The eigenvalues of the matrix A are also called the eigenvalues of the graph. A graph G is said to be connected if and only if there exists a path between any two vertices p and p', i.e., a sequence of vertices and edges $p_0 = p, \ell_1, p_2, \ell_2, \cdots \ell_n, p_n = p'$ where the consecutive vertices and edges are incident. For a graph G without loops, the degree of a vertex is equal to the number of edges incident with the vertex. G is said to be regular if and only if all the degrees are equal. Let G and G' be graphs without multiple edges. G and G' are said to be isomorphic if and only if there exists a bijection $\varphi : V \to V'$ such that v_1 and v_2 are adjacent if and only if $\phi(v_1)$ and

$\phi(v_2)$ are adjacent, $v_1, v_2 \in V$, where V and V' are respectively the set of vertices of G and G'. If G and G' are isomorphic we write $G \cong G'$.

THEOREM (Hoffman and Ray-Chaudhuri [6]). Let v, k and λ be integers and let Π be a symmetric (v, k, λ)-BIBD. Then $G(\Pi)$ is (1) a regular connected graph on vk vertices with distinct eigenvalues -2, $2k-2$, $k-2 \pm \sqrt{k-\lambda}$. Conversely if G is a graph with the properties (1), then either for some symmetric (v, k, λ)-BIBD Π, $G \cong G(\Pi)$ or $(v, k, \lambda) = (4, 3, 2)$ and G is the following graph on 12 vertices.

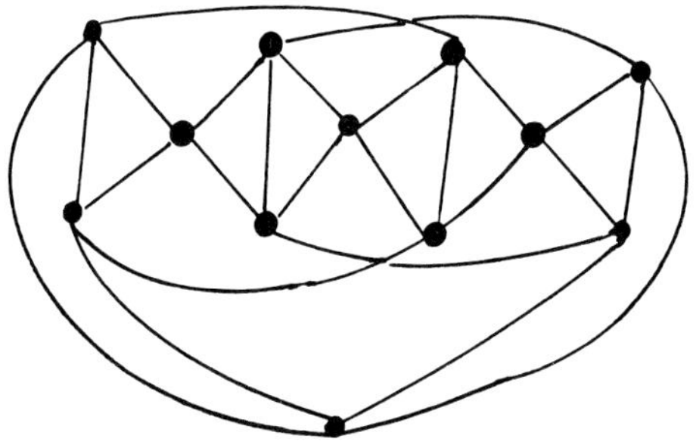

COROLLARY A (v, k, λ)-symmetric BIBD exists if and only if there exists a regular connected graph G on vk vertices with distinct eigenvalues -2, $2k-2$, $k-2 \pm \sqrt{k-\lambda}$. It can be be easily proved that if a (v, k, λ)-BIBD exists, then the triple must satisfy the conditions,

$$\lambda(v-1) \equiv 0 \pmod{k-1}$$

(2)

$$\lambda v(v-1) \equiv 0 \pmod{k(k-1)}$$

and

(3) $$r \geq k.$$

These conditions are known to be necessary but not sufficient. However it is conjectured that these conditions are "asymptotically sufficient" for the existence of a (v, k, λ)-BIBD, i.e., for fixed k and λ, there exists an integer $c(k, \lambda)$ such that if $v \geq c(k, \lambda)$ and v satisfies (2), then a (v, k, λ)-BIBD exists. Recently Richard M. Wilson in his Ph. D. dissertation [14] proved some important results which establish many special cases of this conjecture. Let K be a subset of positive integers. Define $B(K) = \{v: (v, K, 1) - \text{PBD exists}\}$, let $\beta(K)$ be the unique positive integer which generates the ideal generated by the set $\{k(k-1): k \in K\}$ and let $\alpha(K)$ be the unique positive integer which generates the ideal generated by the set $\{(k-1): k \in K\}$. The mapping $K \to B(K)$ of subsets of positive integers into subsets of positive integers is a closure operation i. e.,

(i) $$B(K) \supseteq K$$

(ii) $$K_1 \supseteq K_2 \Rightarrow B(K_1) \supseteq B(K_2)$$

and

(iii) $$B(B(K)) = B(K) .$$

A set K is said to be closed if and only if $K = B(K)$. The existence conjecture about BIBD's is connected with the problem of determining closed sets. A fiber of a closed set K is a residue class mod $\beta(K)$. A fiber d is said to be complete if and only if there exists a constant M such that

$$\{ v: \geq M, \ v \equiv d \ (\mathrm{mod}\ \beta(K)) \} \subseteq K .$$

A closed set K is said to be eventually periodic with period $\beta(K)$ if and only if all fibers of K are complete.

THEOREM (Wilson [14]) All closed sets K are eventually periodic with period $B(K)$.

COROLLARY 1. If k is a prime power, then the necessary conditions (2) are asymptotically sufficient.

COROLLARY 2. If $(k, \lambda) = 1$ or a prime power, then the necessary conditions (2) are asymptotically sufficient.

If a (v, k, λ)-symmetric BIBD, exists, in addition to (2) and (3), the triple (v, k, λ) must satisfy the following:

(4) (a) If v is even, then $(k-\lambda)$ is a square
and
(b) If v is odd, the diophantine equation

$x^2 = (k-\lambda)\, y^2 + (-1)^{\frac{v-1}{2}} \lambda\, z^2$ has a nontrivial solution.

This necessary condition eliminates many infinite families of triples (v, k, λ) which satisfy (2) and (3). The integer c such that p^c divides n but p^{c+1} does not divide n is called the exponent of p in n. One consequence of (4) is that if the exponent of p in n is odd where p is a prime and $p \equiv 3 \pmod 4$, then an $(n^2 + n+1,\ n+1,\ 1)$-BIBD does not exist. The condition (4) for $\lambda = 1$ was proved by Bruck and Ryser whose result was generalized by Shrikhande and Schutzenberger (See Ryser [11], chapter 8). The non-existence of certain asymmetrical (v, k, λ)-BIBD's for triples (v, k, λ) satisfying (2) and (3) is proved by the Hall-Connor embedding theorem (For reference see Ryser [11], p. 128).

Let $(P, \mathfrak{A})$ be a symmetric (v, k, λ)-BIBD. Let $A_{\ell_0} \in \mathfrak{A}$. Then define $P^* = P - A_{\ell_0}$, $A_\ell^* = A_\ell - A_\ell \cap A_{\ell_0}$, $L^* = L - \{\ell_0\}$ and $\mathfrak{A}^* = (A_\ell^*,\ \ell \in L^*)$. $(P^*, \mathfrak{A}^*)$ is a (v^*, k^*, λ^*)-BIBD with $v^* = v - k$, $k^* = k - \lambda$ and $\lambda^* = \lambda$. The BIBD $(P^*, \mathfrak{A}^*)$ will be called a residual BIBD of $(P, \mathfrak{A})$.

THEOREM (Hall and Connor): Let $v = \frac{k(k-1)}{2} + 1$ and $v^* = v - k$, $k^* = k - 2$. Every $(v^*, k^*, 2)$-BIBD is the residual design of a symmetric $(v, k, 2)$-BIBD.

COROLLARY Let $v = \frac{k(k-1)}{2} + 1$, $v^* = v - k$, $k^* = k - 2$. If a $(v^*, k^*, 2)$-BIBD exists, then the triple $(v, k, 2)$ satisfies the conditions (4). From this corollary, it follows that a $(15, 5, 2)$-BIBD does not exist since then the symmetric $(22, 7, 2)$-BIBD would exist and by condition (4) $k - \lambda = 7 - 2 = 5$ must be a perfect square. The Hall-Connor embedding theorem can be derived from a theorem about graphs. Let G be a graph without loops and multiple edges. To construct the line graph $L(G)$, we take one vertex corresponding to every edge of G. Two vertices of $L(G)$ are adjacent if and only if the corresponding edges of G have a common vertex. The graph G is said to be complete if and only if any two distinct vertices are adjacent.

THEOREM (Chang [3]. Hoffman [5], Shrikhande [13]. Let G_n be the complete graph on n vertices.

(i) $L(G_n)$ is a regular connected graph on $\binom{n}{2}$ vertices with distinct eigenvalues -2, 2n-4, and (n-4).

Conversely

(ii) If $n \neq 8$ and H is a regular connected graph on $\binom{n}{2}$ vertices with distinct eigenvalues -2, 2n-4 and n-4, then H is isomorphic to $L(G_n)$.

The Hall-Connor embedding theorem can be derived from the theorem about $L(G_n)$. Given a $(v^*, k^*, 2)$-BIBD, with $k^* = k - 2$, $v^* = v - k$, $v = \frac{k(k-1)}{2} + 1$, define a graph H whose vertices are the blocks of the design and two vertices will be adjacent if and only if the blocks have exactly one treatment in common. The graph H satisfies the conditions of the last theorem.

6. Resolvable balanced incomplete block designs.

Let $(P, \mathfrak{A})$ be a (v, k, λ)_BIBD. A subclass $\mathfrak{A}_1 \subseteq \mathfrak{A}$ is said to be a parallel class of blocks if every treatment occurs in exactly one block of $\mathfrak{A}_1$. The (v, k, λ)-BIBD is said to be resolvable if and only if the family $\mathfrak{A}$ of blocks can be partitioned into r parallel classes $\mathfrak{A}_i$, $i = 1, 2, \cdots, r$. The following is a (9, 3, 1)-resolvable BIBD with P = $\{1, 2, 3, \cdots, 9\}$ and 4 parallel classes $\mathfrak{A}_i$, $i = 1, 2, 3, 4$, each parallel class consisting of 3 blocks.

$\mathfrak{A}_1$	$\mathfrak{A}_2$	$\mathfrak{A}_3$	$\mathfrak{A}_4$
(1, 2, 3)	(1, 4, 7)	(1, 5, 9)	(3, 5, 7)
(4, 5, 6)	(2, 5, 8)	(2, 6, 7)	(2, 4, 9)
(7, 8, 9)	(3, 6, 9)	3, 4, 8)	(1, 6, 8)

The (v, k, λ)-resolvable BIBD is useful for constructing efficient experimental designs when some of the extraneous factors affecting the experimental units are hierarchical in nature. For instance the (9, 3, 1)-resolvable BIBD could be used to construct an experimental design for comparing 9 different ice creams. There will be 12 tasters divided into 4 age groups, each age group containing 3 tasters. Each taster will taste 3 different ice creams successively and give them scores. It is assumed that the score will depend on the quality of the ice cream (the ice cream effect), the age group of the taster (age group effect) and finally the individual taster (the taster effect within the age group). Here the extraneous factor age group is hierarchial in nature. To construct a resolvable BIB experimental design, the 4 age groups are assigned to the 4 parallel classes. Within each age group, the 3 tasters are assigned to the 3 blocks and each taster tastes the 3 ice creams of the assigned block. Here each taster represents a group of 3 experimental units.

If a (v, k, λ)-resolvable BIBD exists, in addition to (2) and (3), the triple (v, k, λ) must satisfy

$$v \equiv 0 \pmod{k} . \tag{5}$$

From (2) and (5), we can show that the existence of a $(v, k, 1)$-resolvable BIBD implies

$$v \equiv k \pmod{k(k-1)}. \tag{6}$$

A $(v, 3, 1)$-resolvable BIBD is equivalent to a Kirkman school girl arrangement for v girls. In 1847, Rev. T. J. Kirkman proposed the following problem of arranging 15 school girls. In a boarding school there are 15 girls and the teacher takes them out for a walk every day of the week. The girls walk in five rows of three. The problem is to find a different row arrangement for the 7 days so as to ensure that any two girls walk in the same row on exactly one day of the week. In general, for v girls, each day the girls are arranged in $\frac{v}{3}$ rows of three. The problem is to find different arrangements for $\frac{v-1}{2}$ consecutive days such that any two girls walk in the same row on exactly one of the $\frac{v-1}{2}$ days. A Kirkman arrangement for v girls is equivalent to a $(v, 3, 1)$-resolvable BIBD, the v girls correspond to v treatments, the rows correspond to blocks and the $\frac{v-1}{2}$ days correspond to $\frac{v-1}{2}$ parallel classes. Hence a necessary condition for the existence of a Kirkman arrangement for v girls is

(7) $$v \equiv 3 \pmod 6 .$$

During the late 19th century and early 20th century, a large number of mathematicians worked on this problem and proved the existence of Kirkman arrangements for many infinite families of integers of the form $6t+3$. However it was not known whether the necessary condition (7) is also sufficient. In 1968, Ray-Chaudhuri and Wilson [9] proved the sufficiency of condition [7].

THEOREM (Ray-Chaudhuri and Wilson [9]). A $(v, 3, 1)$-resolvable BIBD exists if and only if $v \equiv 3 \pmod 6$.
A complete bibliography of the Kirkman's school girl problem can be found in [10]. The necessary condition (6) for the existence of a $(v, k, 1)$-resolvable BIBD is known to be not sufficient. To give an example $(36, 6, 1)$ satisfies (6), but from the Bruck-Ryser theorem, it follows that a $(36, 6, 1)$-resolvable BIBD does not exist. However Ray-Chaudhuri and Wilson [10] recently proved that the condition (6) is asymptotically sufficient.

THEOREM (Ray-Chaudhuri and Wilson [10]) For every integer k, there exists a constant $c(k)$ such that if $v \geq c(k)$ and $v \equiv k \pmod{k(k-1)}$ then a $(v, k, 1)$-resolvable BIBD exists.

It is possible to give a graph theoretical characterization of $(n^2, n, 1)$-resolvable BIBD's. Let $\Pi = (P, L, I)$ be an $(n^2, n, 1)$-resolvable BIBD. Define $g(\Pi)$ as in section 5.

THEOREM (Hoffman and Ray-Chaudhuri [7]). H is a regular connected graph on $n^2(n+1)$ vertices with distinct eigenvalues -2, $2n-1$, $n-2$, $\frac{1}{2}((2n-3) \pm \sqrt{(4n+1)})$ if and only if $H \cong G(\Pi)$, where Π is an $(n^2, n, 1)$-resolvable BIBD.

COROLLARY. An $(n^2, n, 1)$-resolvable BIBD exists if and only if a graph H with properties given in the above theorem exists.

8. Orthogonal array designs.

An (m, n)-orthogonal array is a mapping $A: S_1 \times S_2 \rightarrow S_3$ where S_1 is a set of m "rows", S_2 is a set of n^2 "columns", and S_3 is a set of n symbols, such that for $a, b \in S_1$, $a \neq b$ and $c, d \in S_2$, there exists exactly one column x such that $A(a, x) = c$ and $A(b, x) = d$. The following is a (4, 5) orthogonal array.

1 1 1 1 1	2 2 2 2 2	3 3 3 3 3	4 4 4 4 4	5 5 5 5 5
1 2 3 4 5	1 2 3 4 5	1 2 3 4 5	1 2 3 4 5	1 2 3 4 5
2 3 4 5 1	3 4 5 1 2	4 5 1 2 3	5 1 2 3 4	1 2 3 4 5
3 4 5 1 2	5 1 2 3 4	2 3 4 5 1	4 5 1 2 3	1 2 3 4 5

An (m, n)-orthogonal array can be used to construct an efficient experimental design when there are (m-1) extraneous factors, each having n levels, and n different treatments. There will be n^2 experimental units, each corresponding to one of the n^2 columns. The first (m-1) rows represent the levels of the extraneous factors and the last row represents the treatments applied to the different experimental units. Such an experimental design may be called an (m, n)-OA experimental design. A (3, n)-OA experimental design is the same as a Latin square design and the (4, m)-OA experimental design is the Graeco-Latin square design. Kiefer proved that under the assumption of the additivity of treatment effects and the effects of the extraneous factors, the (3, n)-orthogonal array design has several optimality properties. The general (m, n)-orthogonal array experimental design also possesses similar optimality properties. To give an example, consider an experiment for testing the effects of 5 different types of music on production in a factory. The 5 types of musics are the 5 treatments. It is recognized that production in a factory also depends on the worker, the day of the week, and the week of the month. Hence workers, days, and weeks are considered as extraneous factors and a (4, 5)-OA experimental design is considered suitable. The 4 rows of the (4, 5)-OA will respectively represent the workers, days, weeks, and the type of music. For instance the music type 3 will be played for the second worker on the first day of the first week, if the (4, 5)-OA given above is used.

An (m, n)-OA experimental design is a fractional factorial experiment. The complete factorial experiment will require n^m experimental units whereas the OA experimental design requires only n^2 experimental units. Hence it is desirable to know the largest integer m for which an (m, n)-OA exists. It should be noted that an (m, n)-OA exists if and only if a set of (m-2) mutually orthogonal Latin squares of order n exists. It is easily proved that if an (m, n)-OA exists, then $m \leq n+1$. Conversely Bruck [2] proved the following.

THEOREM (Bruck [2]): If an (m, n)-OA exists, where $n \geq \frac{1}{2}(d-1)(d^3-d^2+d+2)$, $d = n+1-m$, then an (n+1, n)-OA exists.

Bruck's theorem [2] was generalized by Bose's theorem [1] which is best stated as a theorem about graphs. Let G be a graph without loops and multiple edges. Two distinct vertices will be called first associates of each other if and only if they are adjacent. G is said to be a strongly regular graph with parameters $(v, n_1, p_{11}^1, p_{11}^2)$ if and only if

(i) the number of vertices is v, (ii) every vertex has degree n_1, (iii) the number of common first associates of two adjacent vertices is p_{11}^1, and (iv) the number of common first associates of two non-adjacent vertices is p_{11}^2. $\Pi = (P, L, I)$ is said to be an (r, k, t)-partial geometry if and only if (i) every point is incident with r lines, (9 (ii) every line is incident with k points, (iii) two distinct points p_1 and p_2 are incident with at most one common line, and (iv) for every non-incident pair (p, ℓ), there are t lines which are incident with p and also intersect ℓ. Let $G(\Pi)$ denote the graph whose vertex set is P and two vertices p_1 and p_2 in P are adjacent if and only if there is a line ℓ in Π incident with both p_1 and p_2. Let $1 \leq t \leq r, k$. Then Bose [1] proved the following.

THEOREM (Bose [1]). (i) If Π is an (r, k, t)-partial geometry, then $G(\Pi)$ is a $v, n_1, p_{11}^1, p_{11}^2$)-strongly regular graph where $n_1 = r(k-1)$, $v-1-n_1 = \frac{1}{t}(r-1)(k-1)(k-t)$, $p_{11}^1 = (r-1)(k-1) + (k-2)$, and $p_{11}^2 = rt$.

(ii) Conversely if G is a strongly regular graph with parameters as in (i) and $k \geq \frac{1}{2}(r(r-1) + t(r+1)(r^2-2r+2))$, then G is isomorphic to $G(\Pi)$ for an (r, k, t)-partial geometry Π.

Bruck's theorem follows easily from Bose's theorem since an (m, n)-OA exists if and only if (m, n, m-1)-partial geometry exists.

REFERENCES

1. Bose, R. C., "Strongly Regular Graphs, Partial Geometries and Partially Balanced Designs", Pacific J. Math., 13 (1963), 389-419.

2. Bruck, R. H., "Finite Nets II, Uniqueness and Embedding", Pacific J. Math., 13 (1963), 421-457.

3. Chang, L. C., "The Uniqueness and Nonuniqueness of the Triangular Association Schemes", Science Record 3 (1959), 604-613.

4. Hall Jr., Marshall, Combinatorial Theory, Blaisdell Publishing Co., Waltham, Mass., 1967.

5. Hoffman, A. J., "On the Uniqueness of triangular Association Schemes", Ann. Math. Statist. 29, (1958), 262-266.

6. Hoffman, A. J. and Ray-Chaudhuri, D. K., "On the Line Graph of Symmetric Balanced Incomplete Block Designs", Trans. Amer. Math. Soc., 116 (1965), 238-252.

7. Hoffman, A. J. and Ray-Chaudhuri, D. K., "On the Line Graph of a Finite Affine Plane", Can. J. Math., 17 (1965), 687-694.

8. Kiefer, J., "On the Non-Randomized Optimality and Randomized Optimality of Symmetrical Designs", Ann. Math. Statist., 29 (1958), 675-699.

9. Ray-Chaudhuri, D. K. and Wilson, Richard M. "Solution of Kirkman's School Girl Problem", Proceedings of the Symposia in Pure Mathematics, Combinatorics, (19) Amer. Math. Soc.

10. Ray-Chaudhuri, D. K. and Wilson, Richard M., "On the the Existence of Resolvable Balanced Incomplete Block Designs", to appear in the Proceedings of the Calgary International Conference on Combinatorial Structures and their Applications, Gordon and Breach, New York, 1970.

11. Ryser, H. J., Combinatorial Mathematics, Carus Math. Monograph 14, Math. Assoc. Amer., 1963.

12. Smith, C. A. B. and Hartley, H. O., "The Construction of Yonden Squares", J. Roy. Statist. Soc. Ser. B, 10 (1948), 262-263.

13. Shrikhande, S. S., "On a Characterization of the Triangular Association Schemes", Ann. Math. Statist., 30 (1959), 39-47.

14. Wilson, Richard M., "An Existence Theory for Pairwise Balanced Designs", Ph. D. Dissertation, Dept. of Mathematics., The Ohio State University, 1969.

On the Foundations of Combinatorial Theory

GIAN-CARLO ROTA AND RONALD MULLIN

CONTENTS

1. Introduction.

The present work is born from the interplay of two seemingly disparate branches of combinatorial theory. The first is the classical calculus of finite differences, which has been in the past more often related to numerical analysis than to problems of enumeration. In the calculus of finite differences, there occur several sequences of polynomials which are used in interpolation, numerical quadrature, and several other connections. Typical of such sequences of polynomials are the lower factorials

(1.1) $$p_n(x) = (x)_n = x(x-1)\ldots(x-n+1), \quad m = 0, 1, 2, \ldots$$

and the upper factorials

(1.2) $$p_n(x) = x^{(n)} = x(x+1)\ldots(x+n-1), \quad n = 0, 1, 2, \ldots$$

Less well known, but equally significant polynomial sequences are the Abel polynomials, studied by Abel, Hurwitz, Riordan, and others:

(1.3) $$p_n(x) = x(x-an)^{n-1}, \quad n = 0, 1, 2, \ldots$$

and the exponential polynomials, studied by Touchard and others,

(1.4) $$\varphi_n(x) = \sum_{k \geq 0} s(n, k)x^k ,$$

where $s(n, k)$ denote the familiar Stirling numbers of the second kind. Another significant sequence is the Laguerre polynomials

$$L_n(x) = \sum_{k \geq 0} \frac{n!}{k!} \binom{n-1}{k-1}(-x)^k, \tag{1.5}$$

which have an extensive literature. These sequences of polynomials, as well as a large number of other sequences that have arisen in classical analysis and combinatorics, share a common property: that of being of binomial type. We say that a sequence of polynomials $p_n(x)$, where $p_n(x)$ is of exactly of degree of n, is of binomial type when it satisfies the sequence of identities

$$p_n(x+y) = \sum_{k \geq 0} \binom{n}{k} p_k(x) p_{n-k}(y), \quad n = 0, 1, 2, \ldots . \tag{1.6}$$

It will be shown in the course of this study (and it is verified without difficulty using the results below) that each one of the sequences of polynomials mentioned above is of binomial type.

This work is a study of certain analytic or (more suggestively) algebraic-combinatorial properties of sequences of polynomials of binomial type. The main problem we aim at is the following: given two sequences $p_n(x)$ and $q_n(x)$, both of binomial type, there clearly exist coefficients c_{nk}, the so-called connection constants,

$$p_n(x) = \sum_{k \geq 0} c_{nk} q_k(x) \tag{1.7}$$

which express one sequence of polynomials in terms of the other. Our problem is to determine as efficiently as possible the coefficients c_{nk} in terms of minimal data on the polynomials $p_n(x)$ and $q_n(x)$. A few classical instances of this problem are given below.

In trying to solve this problem we were led to develop a systematic theory of polynomial sequences of binomial type. The main novelties appearing in this theory are, first, a systematic use of operator methods as against less efficient generating function methods, which have been used almost exclusively in the past, and secondly a remarkably simple solution of the connection problem stated above, which apparently has eluded past workers in the field.

Bits and pieces of the theory developed in this work can be found in the literature of the last 70 years, starting with the work of Pincherle and Amaldi in 1900, following through the Danish and Italian schools of calculus of finite differences, culminating in the work of I. M. Sheffer and of the great Danish actuarialist Steffensen and in the well-known book by J. Riordan. The statement (though not the proof) of Theorem 4 below is due to Steffensen. A few other results, such as the Expansion Theorem, were at least intuited by Pincherle and his school. But we believe that our notion of umbral operator (a term introduced by Sylvester and extensively used by the invariant theorists and by E. T. Bell, though never correctly defined), together with the solution of the connection constants problem that it yields, gives a new direction to the calculus of finite differences, even for workers interested in purely analytic matters.

It turns out that there is a second and entirely different point of view from which the theory of polynomials of binomial type can be looked at. Each of the polynomial sequences listed above can be interpreted as counting the number of ways of placing "balls" into "boxes", subject to various restrictions. This ties in with the classical theory of distribution and occupancy, which can be alternatively considered as making words out of an alphabet, subject to various restrictions on the successions of letters. More precisely, we are given a set S with n elements and a set X with x elements, and we consider functions from the set S to the set X subject to various restrictions. The restrictions are such that they do not limit the range of the functions but only the domain. Thus, for example, the lower factorial powers (1) count the number of one-to-one functions from a set of n elements to a set of x elements. Similarly, the <u>upper factorials</u>

(1.8) $$x^{(n)} = x(x+1)(x+2)\ldots(x+n-1)$$

count the different ways of placing the balls S into the boxes X when a linear ordering of the balls within each box is to be chosen.

In the same vein, the Abel polynomials

(1.9) $$x(x-an)^{n-1}, \quad n = 0, 1, \ldots, an < x,$$

can be considered in combinatorial terms. Indeed, consider a circle of circumference x, and a set of n arcs each of length a and each having the same radius of curvature as the circle. If we drop the arcs randomly on the circumference of the circle then the probability that no two arcs overlap is easily seen to be

(1.10) $$\frac{x(x-na)^{n-1}}{x^n}.$$

Thus the Abel polynomials "count" the ways (i.e., the measure, since this case is continuous) in which the arcs may be placed without overlapping.

Whenever we count a set of functions from a set of S to a set X, subject to restrictions on the domain, then, letting $p_n(x)$ be the number of such functions, we see immediately that $p_n(x)$ is a polynomial and that the sequence $p_n(x)$ must be of binomial type. Thus, sequences of polynomials of binomial type arise naturally as the unifying concept in the theory of distribution and occupancy.

Accordingly, the present study is divided into two parts. In the first (the present) part we concentrate on the analytic properties of polynomial sequences of binomial type; the relationship to problems of distributions and occupancy is discussed only in Sections 2 and 10, and is meant only as an introduction to the second part. It turns out that every sequence of binomial type with positive integral coefficients can be associated to a counting problem of a certain class of "reluctant" functions, as defined in the next Section. In the second part of this work, which is to follow, we interpret the analytic results derived here in purely combinatorial, that is, set-theoretic terms.

Perhaps the most satisfying results of the present investigation are, first, the unexpected relations of sequences of binomial type with problems of enumeration of rooted labeled trees, (Section 2), and secondly, the solution of the problem of the connection constants, which has deep combinatorial implications.

In several special cases, classical analysis has already answered the problem of the connection constants. For example, we have

$$x^n = \sum_{k\geq 0} S(n,k)(x)_k \tag{1.11}$$

$$(x)_n = \sum_{k\geq 0} s(n,k)x^k \tag{1.12}$$

$$x^{(n)} = \sum_{k\geq 0} |s(n,k)|x^k \tag{1.13}$$

where $s(k,n)$ and $S(k,n)$ are the Stirling number of the first and second kind. Another example is

$$x^{(n)} = \sum_{k\geq 0} \frac{n!}{k!}\binom{n-1}{k-1}(x)_k \tag{1.14}$$

$$L_n(x) = \sum_{k\geq 1} (-1)^k \frac{n!}{k!}\binom{n-1}{k-1}x^k \tag{1.15}$$

where $L_n(x)$ are the Laguerre polynomials. We hope that this introduction has given an idea of the scope of the present investigation. In the next Section we briefly outline some combinatorial connections, thereafter to dismiss them in favor of the analytic theory, until Section 10.

We thank T. Dowling, S. Smoliar and J. Riordan for help in the preparation of this manuscript, and the Statistics Departme of the University of North Carolina, as well as the U. S. Army Research Center at the University of Wisconsin, and Florida Atlantic University for giving us the opportunity to present these results to an actively contributing audience.

2. Reluctant Functions.

Given a function $f: S \to X$, where from now on S will be a finite set with n elements and X will be a finite set of x elements, we can associate with it "functorially" two objects: the range of f, namely, the subset of elements of X which are images of some elements of S under the function f; and the coimage of f, which consists of the partition of the set X defined by the following equivalence relation: an element a of S is equivalent to an element b of S if and only if $f(a) = f(b)$.

We are now going to rather drastically generalize these concepts.

We define a reluctant function from S to X as follows. It is a function f from S to the disjoint union $S \cup X$, subject to the following restriction. For every element $s \in S$, the element $f(f(s))$ is defined if and only if $f(s) \in S$; similarly, $f(f(f(s)))$ is defined if and only if $f(f(s)) \in S$, etc. Our requirement is that only a finite number of terms of the sequence s, $f(s)$, $f(f(s))$, $f(f(f(s)))$, ... be well-defined. A more suggestive, if less precise, way of stating the same condition is the following: for every element $s \in S$, the "orbit" $s, f(s), f(f(s)), f(f(f(s))), \ldots$ of s under iteration of the function f "eventually" ends up in X, where it stops. Thus, one might say that f "reluctantly" maps S into X.

The range of a reluctant function f will consist of those elements of X which are images of some element of S, just like in the case of an ordinary function. On the other hand, we need to generalize the notion of coimage of of an ordinary function, as defined above, to the newly introduced concept of a reluctant function. Whereas the coimage of an ordinary function is simply a partition of the set S, the coimage of a reluctant function is going to be more than a partition of S. In fact, for every element u

of X which is in the range of the reluctant function f, the inverse image of the element u is defined as the set of all elements s of S such that the sequence of its successors f(s), f(f(s)), ... eventually ends up in u. The inverse images of distinct elements of X are disjoint subsets (blocks) of S. Thus, to every reluctant function there is associated a partition of the set S, just like in the case of an ordinary function. However, within each block of such a partition there is a natural structure of a forest of rooted trees describing the "history" of the elements of that block before they end up in X. Thus, we are led to define the coimage of a reluctant function to be a partition of the set S, together with a structure of a rooted forest (i. e., set of rooted trees) defined on each block of the partition. Each rooted forest covering one block of the coimage is the "inverse image" - in the generalized sense just described - of an element x of X.

Note that each block of the coimage can be further partitioned into its connected components, which are all the trees of the rooted forest. The resulting partition is a refinement of the coimage and has the additional property that each block has the structure of a rooted tree. We call this finer partition π of S, together with the structure of rooted tree (See Harary or Moon for definitions) on each block of π, the pre-image of the reluctant function f (recall that a rooted tree is a partially ordered set). Thus, the coimage of f is obtained by "piecing together" all those blocks of the pre-image of f which are "eventually mapped" to the same element x in X.

Clearly, the pre-image of any reluctant function is a rooted labeled forest on the set S, again following classical terminology. Given any rooted labeled forest L on S, with k blocks, that is, consisting of k rooted trees, there are evidently x^k reluctant functions whose pre-image is the forest L.

By way of example, let us consider the set of all reluctant functions from S to X (notice that our use of the word "from" and "to" is not strictly correct, but is nevertheless suggestive so we shall keep using it). Let c_{nk} be the number of rooted labeled forests with k roots on the set S. Then the number of reluctant functions from S to X is evidently given by the polynomial

$$\sum_{k \geq 0} c_{nk} x^k = A_n(x). \tag{2.1}$$

It is easy to see, by a simple combinatorial argument which imitates the standard set-theoretic proof of the binomial theorem, that the sequence of polynomials $A_n(x)$ is of binomial type. It is less obvious, and it will trivially follow from the present theory (see Section 10) that the polynomials $A_n(x)$ are given by the expression

$$A_n(x) = x(x+n)^{n-1}, \tag{2.2}$$

that is, that they are a special case of the Abel polynomials, corresponding to $a = -1$. This gives immediately the classical result of Cayley counting the number of rooted trees, since rooted trees correspond to reluctant functions having as pre-image a partition with one block, and so are the coefficients c_{n1} in (2), which equal n^{n-1}.

We define a <u>binomial class</u> B of reluctant functions as follows. To every set S and set X we assign a set $F(S, X)$ of reluctant functions from S to X. The assignment is "functorial" - or, in combinatorial language, "unlabeled". This means that isomorphisms of the sets S with S_0 and X with X_0 induce a natural isomorphism of the sets $F(S, X)$ with $F(S_0, X_0)$. Thus, if the polynomials $p_n(x)$ denotes the size of the sets $F(S, X)$, the function $p_n(x)$ depends only on the size n of the set S and the size x of the set X.

We come now to the crucial condition. In set-theoretic terms, the condition states that there is a natural isomorphism

$$F(S, X \oplus Y) = \sum_{A \subseteq S} F(A, X) \otimes F(S-A, Y). \tag{2.3}$$

Here, $\oplus$ and $\sum$ denote disjoint sum of sets, $\otimes$ denotes product of sets, and $=$ stands for natural isomorphism. The variable A ranges over all subsets of the set S. We set (for good reasons) $F(\emptyset, X) = 1$ for all non-empty sets X.

Taking the sizes of both sides of (*) we obtain the equation

$$p_n(x+y) = \sum_{k \geq 0} \binom{n}{k} p_k(x) p_{n-k}(y),$$

which expresses the fact that the polynomials $p_n(x)$ are a sequence of binomial type.

Roughly speaking, condition (*) states that by "piecing together" two reluctant functions in the family B, we again obtain a reluctant function in the family. It is a generalized set-theoretic version of the binomial theorem.

Two important ways of defining binomial classes B of reluctant functions are the following. Let T be a family of rooted trees (it is immaterial whether the trees are labeled or unlabeled). The family $B(T)$ will consist of all reluctant functions whose pre-images are labeled forests on S each of whose components is isomorphic to a tree in the family T. Clearly $B(T)$ is a binomial class of reluctant functions. In the example considered above, the family T consisted of all rooted trees.

Thus, we see that the enumeration of labeled forests is closely connected with the theory of polynomials of binomial type. The family T can be specified in innumerable ways, which will be considered in the second part of the present work. For the moment, we shall give some illustrations that show that the classical polynomials listed in Section 1 can be interpreted as enumerating binomial classes of reluctant functions. We have already seen above that the Abel polynomials can be interpreted as enumerating the binomial class of all reluctant functions, at least for $a = -1$. A somewhat more elaborate argument would show that all the other Abel polynomials, for a a negative integer, enumerate other binomial classes of reluctant functions.

Perhaps the simplest example is given by the sequence x^n. This enumerates the binomial class $B(T)$, where T consists of a single tree, with one root.

Another interesting example is the sequence of Laguerre polynomials $L_n(-x)$. These enumerate the binomial class B(T), where T is the set of all linearly ordered rooted trees. We leave the easy verification of this fact to the reader.

A fourth example comes for the inverses of the Abel polynomials, considered in Section 10, namely, functional digraphs, enumerated by the polynomials

$$p_n(x) = \sum_{k \geq 0} \binom{n}{k} k^{n-k} x^k ,$$

which do not appear at first sight to be of binomial type. We prove that they are, by showing that they enumerate a binomial class B(T). Simply take T to be the family of all rooted trees, all of whose branches have length at most two!

Given a binomial class B(T) of reluctant functions, we can consider the subclass of those functions having the property that their coimage coincides with their pre-image. We denote this subclass by $B_m(T)$, and call it the monomorphic class associated with B(T); it generalizes the notion of a one-to-one function.

The monomorphic class associated with x^n consists precisely of all one-to-one functions, enumerated by the lower factorials $(x)_n$. The monomorphic class associated with the Laguerre polynomials turns out to be enumerated by the upper factorials $x^{(n)}$ (as follows from the combinatorial interpretation of $x^{(n)}$ given above).

We state without proof (but the proof is easy) an important result about monomorphic classes. If the sequence

$$p_n(x) = \sum_{k=0}^{n} a_{nk} x^k$$

enumerates the binomial class B(T), then the sequence of polynomials

$$q_n(x) = \sum_{k=0}^{n} a_{nk} (x)_k$$

enumerates the monomorphic class $B_m(T)$. This fact makes formula (14) of the preceeding Section immediately obvious, and a similar interpretation can be given to (11).

The substitution of $(x)_k$ for x^k is an instance of umbral substitution, studied generally in Section 7. It will be seen in the second part of this work that the general umbral substitutions of one basic sequence into another have combinatorial interpretations in terms of "piecing together" trees and other set-theoretic operations.

These examples should suffice to orient the reader to the combinatorial aspect of the theory we are about to develop. The notion of reluctant function does not exhaust the interpretation of sequences of polynomials of binomial type which have negative or non-integral coefficients. Nevertheless, we shall see in the second part of this work that all sequences of polynomial type with non-negative coefficients can be set-theoretically (or probabilistically) interpreted by a generalization of the notion of reluctant functions, whereas those with negative coefficients can be interpreted by sieving methods (Möbius inversions, etc.). There is also an obvious connection with the theory of compound Poisson processes.

Apologizing for this sketchy introduction, we proceed to begin the analytic theory.

3. Fundamentals.

Throughout this paper, we shall be concerned with the algebra (over a field of characteristic zero) of all polynomials in one variable, to be denoted $\underline{P}$.

By a polynomial sequence we shall denote a sequence of polynomials $p_i(x)$, $i = 0, 1, 2, \ldots$ where $p_i(x)$ is exactly of degree i, for all i.

A polynomial sequence is said to be of binomial type if it satisfies the infinite sequence of identities

$$p_n(x+y) = \sum_{k\geq 0} \binom{n}{k} p_k(x) p_{n-k}(y), \quad n = 0, 1, 2, \ldots .$$

All the polynomial sequences mentioned above are of binomial type. For some sequences, such as x^n, this is a trivial observation, but for others, such as the Abel and Touchard polynomials, the verification that they are of binomial type will be a consequence (a rather simple one, to be sure) of our theory.

Our study will revolve primarily around the study of linear operators on $\underline{P}$ considered as a vector space. Henceforth, all operators we consider will be tacitly assumed to be linear. We denote the action of an operator T on the polynomial $p(x)$ by $Tp(x)$, this notation is not strictly correct; a correct version is $(Tp)(x)$. However, this notational license results in greater readability. By way of orientation, we list some of the operators of frequent occurrence in the theory of binomial enumeration. The most important are the shift operators. A shift operator, written E^a, is an operator which translates the argument of a polynomial by a, where a is an element of the field, that is, $E^a p(x) = p(x+a)$.

An operator T which commutes with all shift operators is called a shift-invariant operator, i.e.,

$$TE^a = E^a T .$$

The following are important examples of shift-invariant operators:

(i) Identity operator $I: x^n \to x^n$.

(ii) Differentiation operator $D: x^n \to nx^{n-1}$.

(iii) Difference operator $\Delta = E - I: (x)_n \to n(x)_{n-1}$, where we write E in place of E^1, where 1 is the identity of the field.

(iv) The Abel operator $DE^a = E^a D: x(x-na)^{n-1} \to nx(x-(n-1)a)^{n-2}$.

(v) Bernoulli operator $J: p(x) \to \int_x^{x+1} p(t)dt$.

(vi) Backward difference operator

$$\nabla = I - E^{-1}: x^{(n)} \to nx^{(n-1)}.$$

(vii) Laguerre operator

$$L: p(x) \to \int_0^{\infty} e^{-t} p'(x+t)dt .$$

(viii) Hermite operator

$$H: p(x) \to \sqrt{\frac{2}{\pi}} \int_{-\infty}^{\infty} e^{-t^2/2} p(x+t)dt.$$

(ix) Central difference operator

$$\delta = E^{1/2} - E^{-1/2} : p(x) \to p(x+1/2) - p(x-1/2).$$

(x) Euler (mean) operator $M = (1/2)(I+E)$: $p(x) \to (1/2)(p(x) + P(x+1))$.

We define a delta operator, (a term suggested by F. B. Hildebrandt) usually denoted by the letter Q, as a shift-invariant operator for which Qx is a non-zero constant.

The derivative, difference, backward difference, central difference, Laguerre, and Abel operators are delta operators.

Delta operators possess many of the properties of the derivative operator, as we proceed to show.

Lemma 1: If Q is a delta operator, then $Qa = 0$ for every constant a.

Proof: Since Q is shift invariant, then

$$QE^a x = E^a Qx .$$

By the linearity of Q,

$$QE^a x = Q(x+a) = Qx+Qa = c+Qa,$$

since Qx is equal to some non-zero constant c by definition. But also

$$E^a Qx = E^a c = c$$

and so $c + Qa = c$. Hence $Qa = 0$, Q. E. D.

Lemma 2: If $p(x)$ is a polynomial of degree n and Q is a delta operator, then $Qp(x)$ is a polynomial of degree $n-1$.

Proof: It is sufficient to prove the conclusion for the special case $p(x) = x^n$, that is, to show that the polynomial $r(x) = Qx^n$ is of degree $n-1$ (exactly). From the binomial theorem and the linearity of Q we have

$$Q(x+a)^n = \sum_{k \geq 0} \binom{n}{k} a^k Q x^{n-k} .$$

Also by the shift invariance of Q

$$Q(x+a)^n = QE^a x^n = E^a Q x^n = r(x+a)$$

so that

$$r(x+a) = \sum_{k \geq 0} \binom{n}{k} a^k Q x^{n-k} .$$

Putting $x = 0$, we have r expressed as a polynomial in a:

$$r(a) = \sum_{k \geq 0} \binom{n}{k} a^k [Qx^{n-k}]_{x=0} .$$

The coefficient of a^n is

$$[Qx^{n-n}]_{n=0} = [Q1]_{x=0} = 0$$

by Lemma 1. Further, the coefficient of a^{n-1} is

$$\binom{n}{n-1} [Qx^{n-n+1}]_{x=0} = n[Qx]_{x=0} = nc \neq 0 .$$

Hence r is of degree $n-1$, Q. E. D.

Let Q be a delta operator. A polynomial sequence $p_n(x)$ is called the sequence of basic polynomials for Q if:

(1) $p_0(x) = 1$

(2) $p_n(0) = 0$ whenever $n > 0$

(3) $Qp_n(x) = np_{n-1}(x)$.

Using Lemma 2, it is easily shown by induction that every delta operator has a unique sequence of basic polynomials associated with it. For example, the basic polynomials for the derivative operator are x^n .

We shall now see that several properties of the polynomial sequence x^n can be generalized to an arbitrary sequence of basic polynomials. The first property we noticed about x^n was that it was of binomial type. This turns out to be true for every sequence of basic polynomials, and is one of our basic results.

Theorem 1.

(a) If $p_n(x)$ is a basic sequence for some delta operator Q , then it is a sequence of polynomials of binomial type.

(b) If $p_n(x)$ is a sequence of polynomials of binomial type, then it is a basic sequence for some delta operator.

Proof:

(a) Iterating property (3) of basic polynomials, we see that

$$Q^k p_n(x) = (n)_k p_{n-k}(x)$$

and hence that for $k = n$,

$$[Q^n p_n(x)]_{x=0} = n!$$

while

$$[Q^k p_n(x)]_{x=0} = 0 , \quad k < n .$$

Thus, we may express $p_n(x)$ in the following form:

$$p_n(x) = \sum_{k \geq 0} \frac{p_k(x)}{k!} [Q^k p_n(x)]_{x=0} .$$

Since any polynomial $p(x)$ is a linear combination of the basic polynomials $p_n(x)$, this expression also holds for all polynomials $p(x)$, i. e.,

$$p(x) = \sum_{k \geq 0} \frac{p_k(x)}{k!} [Q^k p(x)]_{x=0} .$$

Now suppose $p(x)$ is the polynomial $p_n(x+y)$. Then

$$p_n(x+y) = \sum_{k \geq 0} \frac{p_k(x)}{k!} [Q^k p_n(x+y)]_{x=0} .$$

But

$$\begin{aligned}
[Q^k p_n(x+y)]_{x=0} &= [Q^k E^y p_n(x)]_{x=0} \\
&= [E^y Q^k p_n(x)]_{x=0} \\
&= [E^y (n)_k p_{n-k}(x)]_{x=0} \\
&= (n)_k p_{n-k}(y)
\end{aligned}$$

and so

$$p_n(x+y) = \sum_{k \geq 0} \binom{n}{k} p_k(x) p_{n-k}(y)$$

which means that $p_n(x)$ is of binomial type.

(b) Conversely, suppose $p_n(x)$ is a sequence of binomial type. Putting $y = 0$ in the binomial identity, we have

$$\begin{aligned}
p_n(x) &= \sum_{k \geq 0} \binom{n}{k} p_k(x) p_{n-k}(0) \\
&= p_n(x) p_0(0) + n p_{n-1}(x) p_1(0) + \dots \quad .
\end{aligned}$$

Since each $p_i(x)$ is exactly of degree i, it follows that $p_0(0) = 1$ (and hence $p_0(x) = 1$) and $p_i(0) = 0$ for all other i. Thus properties (1) and (2) of basic sequences are satisfied.

We now find a delta operator for which such a sequence $p_n(x)$ is the sequence of basic polynomials. Let Q be the operator defined by the property that $Qp_0(x) = 0$ and $Qp_n(x) = np_{n-1}(x)$ for $n \geq 1$. Clearly Qx must be a non-zero constant. Hence all that remains to be shown is that Q is shift-invariant.

As before we may trivially rewrite the generalized binomial theorem in terms of Q:

$$p_n(x+y) = \sum_{k \geq 0} \frac{p_k(x)}{k!} Q^k p_n(y)$$

and, by linearity, this may be extended to all polynomials:

$$p(x+y) = \sum_{k \geq 0} \frac{p_k(x)}{k!} Q^k p(y) .$$

Now replace p by Qp and interchange x and y on the right to get

$$(Qp)(x+y) = \sum_{k \geq 0} \frac{p_k(y)}{k!} Q^{k+1} p(x) .$$

But

$$(Qp)(x+y) = E^y(Qp)(x) = E^y Qp(x)$$

and

$$\sum_{k \geq 0} \frac{p_k(y)}{k!} Q^{k+1} p(x) = Q[\sum_{k \geq 0} \frac{p_k(y)}{k!} Q^k p(x)]$$

$$= Q(p(x+y))$$

$$= QE^y p(x) .$$

Thus we have

$$E^y Q p(x) = Q E^y p(x) ,$$

for all polynomials $p(x)$, i.e., Q is shift-invariant,
Q. E. D.

4. Expansions.

We shall study next the various ways of expressing a shift-invariant operator in terms of a delta operator and its powers. The difficulties caused by convergence questions are minimal, and we shall get around them in the easy way.

Consider a sequence of shift-invariant operators T_n on $\underline{P}$. We say that the sequence converges to T, written $T_n \to T$, if for every polynomial $p(x)$ the sequence of polynomials $T_n p(x)$ converges pointwise to the polynomial $Tp(x)$. The convergence of an infinite series of operators is to be understood accordingly.

The following theorem generalizes the Taylor expansion theorem to arbitrary delta operators and basic polynomials.

Theorem 2 (Expansion Theorem). Let T be a shift-invariant operator, and let Q be a delta operator with basic set $p_n(x)$. Then

$$T = \sum_{k \geq 0} \frac{a_k}{k!} Q^k$$

where

$$a_k = [Tp_k(x)]_{x=0} .$$

Proof: Since the polynomials $p_n(x)$ are of binomial type then, as usual, we rewrite the binomial formula as

$$p_n(x+y) = \sum_{k \geq 0} \frac{p_k(y)}{k!} Q^k p_n(x) .$$

Now we may regard this as a polynomial in the variable y and apply T to both sides to get:

$$Tp_n(x+y) = \sum_{k \geq 0} \frac{Tp_k(y)}{k!} Q^k p_n(x)$$

Again, by linearity, this expression can be extended to all polynomials p. After doing this and setting y equal to zero we get

$$Tp(x) = \sum_{k \geq 0} \frac{[Tp_k(y)]_{y=0}}{k!} Q^k p(x) \qquad \text{Q. E. D.}$$

Obviously, the best-known example of this Theorem is when $T = E^a$ and $Q = D$; then $p_n(x) = x^n$, and we have Taylor's expansion. A second example is Newton's expansion, which has three forms. If $Q = \Delta$, then $p_n(x) = (x)_k$ and the coefficients are $a_k = [T(x)_k]_{x=0}$. If $Q = \nabla$ then $p_n(x) = x^{(n)}$ and $a_k = [Tx^{(k)}]_{x=0}$. The basic polynomials for $Q = \delta = E^{1/2} - E^{-1/2}$ will be determined later.

The following remark will be used occasionally:

Lemma: If Q is a delta operator, and $p(x)$, $q(x)$ any polynomials, then

$$[p(Q)q(x)]_{x=0} = [q(Q)p(x)]_{x=0}.$$

Proof: By linearity, we need only consider the cases when $q(x) = p_k(x)$ and $p(Q) = Q^n$, where $p_k(x)$ are the basic polynomials of Q. But it is easy to see that the relation holds in this case. Q. E. D.

As a further example of the use of the expansion theorem, we derive the classical Newton-Cotes formulas of numerical integration. We wish to find an expansion, in terms of Δ, of the Bernoulli operator J_r defined by:

$$J_r p(x) = \int_x^{x+r} p(t)\,dt .$$

Noting that J_r is shift-invariant, we have the identities

$$J_r = \frac{(i+\Delta)^r - I}{\Delta} \cdot \frac{\Delta}{D}$$

$$= \frac{(I+\Delta)^r - I}{\Delta} \cdot J$$

which reduces the problem to finding an expansion of J in terms of Δ. Using the First Expansion Theorem, this is fairly simple:

$$J = \sum_{k \geq 0} \frac{a_k}{k!} \Delta^k$$

where

$$a_k = [J(x)_k]_{x=0} = \int_0^1 (x)_k dx$$

where we note that the a_k are the Bernoulli numbers of the second kind. J_2 evaluated in this way gives Simpson's rule:

$$\int_x^{x+2h} p(t)dt = 2h(1+\Delta+\frac{1}{6}\Delta^2+\frac{1}{180}\Delta^4+\frac{1}{180}\Delta^6+\ldots)p(x) .$$

A final example is the classical Euler's transformation

$$\sum_{k \geq 0} (-1)^k f(k) = 1/2 \sum_{k \geq 0} \frac{(-1)^n}{2^n} \Delta^n f(0)$$

which follows from the identities:

$$\sum_{k \geq 0} (-1)^k E^k = \frac{I}{I+E} = \frac{I}{2I+\Delta}$$

$$= 1/2 \frac{I}{I + 1/2\,\Delta} = 1/2 \sum_{n \geq 0} \frac{(-1)^n}{2^n} \Delta^n .$$

Of course, in this case we are disregarding convergence questions.

We now turn our attention to the Abel polynomials. The delta operator in this case is E^aD. Thus, the Abel polynomials are basic polynomials and hence are of binomial type. Therefore, by Theorem 1 we have proved Abel's identity:

$$(x+y)(x+y-na)^{n-1} = \sum_{k\geq 0} \binom{n}{k} x(x-ka)^{k-1} y(y-(n-k)a)^{n-k-1},$$

not easily proved by direct methods. We can use the Expansion Theorem to get an Abel expansion of e^x. Indeed, we do get the following beautiful expansion

$$e^x = \sum_{k\geq 0} \frac{x(x-ka)^{k-1}}{k!} e^{ka},$$

convergent for $a < 0$.

Theorem 3. Let Q be a delta operator, and let F be the ring of formal power series in the variable t, over the same field. Then there exists an isomorphism from F onto the ring $\sum$ of shift-invariant operators, which carries

$$f(t) = \sum_{k\geq 0} \frac{a_k}{k!} \quad \text{into} \quad \sum_{k\geq 0} \frac{a_k}{k!} Q^k.$$

Proof: The mapping is already linear and by the Expansion Theorem, it is onto. Therefore, all we have to verify is that the map preserves products. Let T be the shift-invariant operator corresponding to the formal power series $f(t)$ and let S be the shift-invariant operator corresponding to

$$g(t) = \sum_{k\geq 0} \frac{b_k}{k!} t^k.$$

We must verify that

$$[TSp_n(x)]_{x=0} = \sum_{k\geq 0} \binom{n}{k} a_k b_{n-k}$$

where $p_n(x)$ are the basic polynomials of Q. Now

$$[TSp_r(x)]_{x=0} = [(\sum_{k\geq 0} \frac{a_k}{k!} Q^k \sum_{n\geq 0} \frac{b_n}{n!} Q^n) p_r(x)]_{x=0}$$

$$= [\sum_{k\geq 0} \sum_{n\geq 0} \frac{a_k b_n}{k!\, n!} Q^{k+n} p_r(x)]_{x=0} .$$

But $p_n(0) = 0$ for $n > 0$ and $p_0(x) = 1$. Hence, it follows that the only non-zero terms of the double sum occur when $n = r-k$. Thus

$$[TSp_r(x)]_{x=0} = [\sum_{k\geq 0} \frac{a_k b_{r-k}}{k!\,(r-k)!} Q^r p_r(x)]_{x=0}$$

$$= [\sum_{k\geq 0} \frac{a_k b_{r-k}}{k!\,(r-k)!} r!\, p_0(x)]_{x=0}$$

$$= \sum_{k\geq 0} \binom{r}{k} a_k b_{r-k} , \qquad \text{Q. E. D.}$$

Corollary 1. A shift-invariant operator T is invertible if and only if $T1 \neq 0$.

In the following, we shall write $P = p(Q)$, where P is a shift-invariant operator and $p(t)$ is a formal power series, to indicate that the operator P corresponds to the formal power series $p(t)$ under the isomorphism of Theorem 3. Note that $p(0) = 0$ and $p'(0) \neq 0$ whenever P is a shift-invariant delta operator. For such formal power series, a unique inverse formal power series $p^{-1}(t)$ exists.

Corollary 2. Let Q be a delta operator with basic polynomials $p_n(x)$, and let $q(D) = Q$. Let $q^{-1}(t)$ be the inverse formal power series. Then

$$\sum_{n\leq 0} \frac{p_n(x)}{n!} u^n = e^{xq^{-1}(u)} .$$

Proof: Expand E^a in terms of Q. The coefficient a^n are $p_n(a)$. Hence

$$\sum_{n\geq 0} \frac{p_n(a)}{n!} Q^n = E^a ,$$

a formula which can be considered as a generalization of Taylor's formula, and which specializes (for example, for $Q = \Delta$ it gives Newton's expansion) to several other classical expansions. Now use the Isomorphism Theorem, with D as the delta operator. We get

$$\sum_{n\geq 0} \frac{p_n(a)}{n!} q(t)^n = e^{at} ,$$

whence the conclusion, upon setting $u = q(t)$ and $a = x$, Q. E. D.

As an aside, we remark at this point a possibly useful connection between basic polynomials and orthogonal polynomials:

Proposition. Let $p_n(x)$, $n = 0, 1, 2, \ldots$ be a sequence of polynomials of binomial type. Then there exists a unique inner product $(p(x), q(x))$, on the vector space $\underline{P}$ of all polynomials $p(x)$, under which the sequence $p_n(x)$ is an orthogonal sequence and $(p_n(x), p_n(x)) = n!$. Under this inner product we have

$$[Q^n p(x)]_{x=0} = (p(x), p_n(x))$$

so that

$$p(x) = \sum_{n\geq 0}^{\infty} \frac{p_n(x)}{n!} [Q^n p(x)]_{x=0} = \sum_{n\geq 0} \frac{p_n(x)}{n!} (p(x), p_n(x)).$$

Proof: Let T be the (uniquely defined) operator mapping $p_n(x)$ to x^n, for all n. Define the inner product as follows:

$$(p(x), q(x)) = [(Tp)(Q)q(x)]_{x=0} .$$

An argument similar to the proof of the Lemma preceding Theorem 2 shows that this bilinear form is symmetric (set $p(x) = p_n(x)$ and $q(x) = p_k(x)$), and that $p_n(x)$ is orthogonal to $p_k(x)$ for $k \neq n$. Finally,

$$(p_n(x), p_n(x)) = [(Tp_n)(Q)p_n(x)]_{x=0} =$$

$$[Q^n p_n(x)]_{x=0} = n! ,$$

which shows that the bilinear form is positive definite.

It is trivially verified that $[Q^n p(x)]_{x=0} = (p(x), p_n(x))$. Thus the Expansion Theorem, in the form

$$q(a) = [E^a q(x)]_{x=0} = \sum_{n \geq 0} \frac{q_n(a)}{n!} [Q^n q(x)]_{x=0}$$

is the same as the orthogonal expansion of $q(x)$ relative to the above inner product, Q. E. D.

We note that for the Laguerre polynomials, discussed below, the inner product just introduced reduces to the classical inner product making the Laguerre polynomials an orthogonal set.

Note that for the operators (i), (ii), (iii), (iv), (vi), (vii), (ix) described at the beginning of this Section the polynomials defined there are the basic sets, as will be shown in the course of this study.

5. Closed Forms.

We now introduce a class of linear operators of an altogether different kind. Let $p(x)$ be a polynomial in the parameter x. Multiplying each term of $p(x)$ by a factor x, i. e., replacing each occurrence of x^n by x^{n+1}, $n \geq 0$, we obtain a new polynomial in x which we may denote $xp(x)$. The first x in this expression may be regarded as a linear operator since it represents a linear transformation of polynomial into polynomials. We call this the multiplication operator and we denote it by the parameter $\underline{x}$ underlined. Thus, $\underline{x}$: $p(x) \to xp(x)$. Note that the operator $\underline{x}$ is not shift-invariant.

Before proceeding further, it should be noted that $E^a p(x) = p(x+a)$ is a polynomial in the formal parameter x+a. Since the multiplication operator is not shift-invariant, we have the operator identity:

$$E^a \underline{x} = (\underline{x+a}) E^a ,$$

where $\underline{x+a}$: $p(x) \to (x+a)p(x)$.

Proposition 1. If T is a shift-invariant operator, then

$$T' = T\underline{x} - \underline{x}T$$

is also a shift-invariant operator.

The proof is a straightforward verification. We call T' the Pincherle derivative of the operator T .

We saw in the previous Section, as a special case of the Expansion Theorem, that any shift-invariant operator T can be expressed as an expansion in the delta operator D, i. e., $T = \sum_{k \geq 0} \frac{a_k}{k!} D^k$ where $a_k = [Tx^k]_{x=0}$. Further, by the isomorphism theorem, (Theorem 3) the formal power series corresponding to T is $\sum_{k \geq 0} \frac{a_k}{k!} t^k = f(t)$. We call f(t) the indicator of T .

Proposition 2. If T has indicator f(t), then T' has $f'(t)$ as its indicator.

Proof: Straightforward verification of coefficients by Theorem 3.

We note in passing Pincherle's Formula:

$$Tx^n p(x) = \sum_{k \geq 0} \binom{n}{k} x^{n-k} T^{(k)} p(x) .$$

Note that by the isomorphism theorem of the preceeding Section, we also have

$$(TS)' = T'S + TS' .$$

Proposition 3. Q is a delta operator if and only if $Q = DP$ for some shift-invariant operator P, where P^{-1} exists.

Proof: If Q is a delta operator, then it can be written

$$Q = \sum_{k \geq 0} \frac{a_k}{k!} D^k$$

where

$$a_k = [Qx^k]_{x=0} .$$

But

$$a_0 = [Q1]_{x=0} = 0$$

$$a_1 = [Qx]_{x=0} \neq 0$$

by definition of a delta operator. Thus if we set

$$P = \sum_{k \geq 0} \frac{a_{k+1}}{(k+1)!} D^k$$

then the conclusion follows at once.

Conversely, suppose $Q = DP$ where P is shift-invariant and P^{-1} exists. Since D and P are shift-invariant, then Q must be also. Further, shift-invariant operators commute (by Theorem 3), so that

$$Qx = DPx = PDx = P1 \neq 0 ,$$

since $P1 \neq 0$ for an invertible shift-invariant operator. Hence Q is a delta operator. Q. E. D.

Theorem 4 (Closed forms for basic polynomials). If $p_n(x)$ is a sequence of basic polynomials for the delta operator $Q = DP$, then

(1) $p_n(x) = Q'P^{-n-1}x^n$

(2) $p_n(x) = P^{-n}x^n - (P^{-n})'x^{n-1}$

(3) $p_n(x) = xP^{-n}x^{n-1}$

(4) (Rodrigues-type formula) $p_n(x)=x(Q')^{-1}p_{n-1}(x)$.

Proof: We shall first show that (1) and (2) define the same polynomial sequence:

$$Q'P^{-n-1} = (DP)'P^{-n-1}$$
$$= (D'P + DP')P^{-n-1}.$$

It is easy to see that $D' = I$. Hence

$$Q'P^{-n-1} = P^{-n}+P'P^{-n-1}D$$
$$= P^{-n} - \frac{1}{n}(P^{-n})'D.$$

Hence,

$$Q'P^{-n-1}x^n = P^{-n}x^n - (P^{-n})'x^{n-1}$$

so that (1) and (2) are equivalent. Recalling the definition of the Pincherle derivative of $(P^{-n})'$, we have

$$P^{-n}x^n - (P^{-n})'x^{n-1} = P^{-n}x^n - (P^{-n}\underline{x}-\underline{x}P^{-n})x^{n-1}$$
$$= xP^{-n}x^{n-1}$$

and thus formula (3) is equivalent to formulas (2) and (1). Now, if we set

$$q_n(x) = Q'P^{-n-1}x^n$$

then by writing $Q = DP$ we get

$$Qq_n(x) = DPQ'P^{-n-1}x^n = Q'P^{-n}Dx^n = nq_{n-1}(x) .$$

Thus, if we can show that $q_n(0) - 0$ for $n > 0$, we will complete the proof that $q_n(x)$ is the sequence of basic polynomials for Q, and it will follow that they will satisfy formulas (1), (2), and (3). Now, from the equivalence of equations (1), (2), (3) we see that

$$q_n(x) = xP^{-n}x^{n-1}$$

and hence $q_n(0) = 0$ for $n \geq 0$. Thus (1), (2), and (3) have been proven, and $q_n(x) = p_n(x)$.

To prove (4), we first invert formula (1), getting:

$$x^n = (Q')^{-1}P^{n+1}p_n(x).$$

Notice that Q' is invertible, as is easily verified. Inserting this into the right side of formula (3) we get:

$$\begin{aligned} p_n(x) &= xP^{-n}(Q')^{-1}P^n p_{n-1}(x) \\ &= x(Q')^{-1}p_{n-1}(x) \end{aligned}$$

which is the Rodrigues-type formula, Q. E. D.

The following formulas, numbered (5) and (6), relate the basic polynomials of two different delta operators in an analogous way. Their proof is immediate.

Corollary. Let $R = DS$ and $Q = DP$ be delta operators with basic polynomials $r_n(x)$ and $p_n(x)$, respectively, where S^{-1} and P^{-1} exist. Then:

(5) $$p_n(x) = Q'(R')^{-1}P^{-n-1}S^{n+1}r_n(x)$$

(6) $$p_n(x) = x(RQ^{-1})^n x^{-1} r_n(x) .$$

Example 1. The Abel polynomials are the basic polynomials of the Abel operator E^aD. Indeed from formula (3):

$$p_n(x) = xE^{-an}x^{n-1}$$
$$= x(x-an)^{n-1} .$$

Example 2. The lower factorials $(x)_n$ are the sequence of basic polynomials for the lower difference operator $\Delta = E-I$. Since $\Delta' = E$, the Rodrigues formula (4) gives immediately

$$p_n(x) = xE^{-1}p_{n-1}(x)$$

which by iteration gives the lower factorial power $(x)_n$.

Note that the basic polynomials for the central difference operator δ can be obtained from (6) and Δ much as the Abel polynomial were obtained from (3).

6. The Automorphism Theorem.

Let $\mathcal{A}(\underline{P})$ be the algebra of all linear operators on the algebra of all polynomials $\underline{P}$. Let Σ be the subalgebra of shift-invariant operators on $\underline{P}$. We now prove our main result.

Theorem 5. Let T be an operator in $\mathcal{A}(\underline{P})$, not necessarily shift-invariant. Let P and Q be delta operators with basic polynomials $p_n(x)$ and $q_n(x)$, respectively. Assume that

$$Tp_n(x) = q_n(x), \quad \text{for all } n \geq 0 ,$$

then T^{-1} exists and

(a) the map $S \to TST^{-1}$ is an automorphism of the algebra Σ.

(b) T maps every sequence of basic polynomials into a sequence of basic polynomials.

(c) Let $P = p(D)$ and $Q = q(D)$, where $p(t)$ and $q(t)$ are formal power series. Let the delta operator R have formal power series expansion $r(D)$ and basic polynomials $r_n(x)$. Then

$$Tr_n(x) = s_n(x)$$

is a sequence of basic polynomials for the delta operator

$$S = r(p^{-1}(q(D))) ,$$

where p^{-1} is the inverse formal power series of $p(t)$, that is $p(p^{-1}(t)) = p^{-1}(p(t)) = t$.

Proof:
(a) We have the string of identities:

$$\begin{aligned} TPp_n(x) &= T(np_{n-1}(x)) \\ &= nTp_{n-1}(x) \\ &= nq_{n-1}(x) \\ &= Qq_n(x) \\ &= QTp_n(x) \end{aligned}$$

and since every polynomial is a linear combination of the basic polynomials, by linearity, we infer that $TPp(x) = QTp(x)$ for all polynomials $p(x)$, that is, $TP = QT$. It is clear that T is invertible, since it maps polynomials of degree n into polynomials of degree n, for all n. Hence

$$TPT^{-1} = Q$$

whence

$$TP^nT^{-1} = Q^n$$

for all $n > 0$. Let S be any shift-invariant operator and let the expansion of S in terms of P be

$$S = \sum_{n \geq 0} \frac{a_n}{n!} P^n .$$

Then

$$TST^{-1} = T\Big(\sum_{n\geq 0} \frac{a_n}{n!} P^n\Big)T^{-1} = \sum_{n\geq 0} \frac{a_n}{n!} Q^n \qquad \text{(I)}$$

and thus TST^{-1} is a shift-invariant operator. Furthermore the map $S \to TST^{-1}$ is onto since any shift-invariant operator can be expanded in terms of Q. Thus, the map is an automorphism, as claimed.

Remark: We have also shown that T maps delta operators into delta operators, since for delta operators the constant co-efficient a_0 vanishes.

(b) Let $s_n(x) = Tr_n(x)$ and let $S = TRT^{-1}$. By the remark above, S is a delta operator since R is. Also

$$\begin{aligned} Ss_n(x) &= TRT^{-1}s_n(x) \\ &= TRr_n(x) \\ &= nTr_{n-1}(x) \\ &= ns_{n-1}(x) . \end{aligned}$$

To complete the proof that $s_n(x)$ are the basic polynomials of S we need only show that $s_n(0) = 0$ for $n > 0$. Now we can write

$$r_n(x) = \sum_{k\geq 1} a_k p_k(x)$$

since $a_0 = 0$ because R is a delta operator, and hence $r_n(0) = 0$. Hence

$$Tr_n(x) = \sum_{k\geq 1} a_k q_k(x) = s_n(x)$$

so that

$$s_n(0) = 0 , \quad n > 0 ,$$

as desired.

(c) Now Q and R can be written as power series in P, say $Q = f(P)$ and $R = q(P)$. In equation (I) above let

$$R = g(P) = \sum_{k \geq 0} \frac{a_n}{n!} P^n ;$$

then

$$S = TRT^{-1} = g(Q)$$

and therefore

$$R = g(P) = g(p(D))$$

and

$$S = g(Q) = g(q(D)).$$

Finally we see that

$$r(D) = g(p(D))$$

$$f(D) = r(p^{-1}(D))$$

and

$$S = g(Q) = r(p^{-1}(q(D))), \qquad \text{Q. E. D.}$$

7. Umbral Notation.

In order to simplify the complex notation which has been appearing in many of the above formulas, we will make use and for the first time make rigorous the "umbral calculus" or "symbolic notation" first devised by Sylvester and later used informally by many authors. If $\{a_n(x)\}$ is a polynomial sequence then we simply note that there is a unique linear operator L on $\underline{P}$ such that $L(x^n) = a_n(x)$. We say that L is the umbral representation of the sequence $\{a_n(x)\}$. In particular, an operator T with the properties specified in the preceding Theorem will be called an umbral operator.

If $f(x)$ is a polynomial then we use the notation $f(\underline{a}(x))$ to denote the image of $f(x)$ under the operator L. For example, $\underline{a}(x)$ denotes $a_1(x)$, while $[\underline{a}(x)]^2$ denotes $a_2(x)$. Similarly, $[\underline{a}(x)+b][\underline{a}(x)+c]$ denotes $a_2(x)+(b+c)a_1(x) + bc\, a_0(x)$. This is in essence the umbral notation, which we signify by boldface lettering.

Loosely speaking, umbral notation is a simple technique for using exponents to denote subscripts. For example, the defining property for a polynomial sequence to be of binomial type

$$p_n(x+y) = \sum_{k\geq 0} \binom{n}{k} p_k(x) p_{n-k}(y)$$

can be restated umbrally as

$$\underline{p}^n(x+y) = [\underline{p}(x) + \underline{p}(y)]^n .$$

Note that, in view of our definition in terms of the operator L, this identity has a well-defined meaning.

Theorem 6. If P and Q are delta operators with basic sequences $p_n(x)$ and $q_n(x)$, and expansions $P = p(D)$ and $Q = q(D)$, then the umbral composition

$$r_n(x) = p_n(\underline{q}(x))$$

is the sequence of basic polynomials for the delta operator

$$R = p(q(D)).$$

Proof: Let T be the umbral operator defined by

$$Tx^n = q_n(x) .$$

By the Automorphism Theorem of the preceding Section, it follows that T takes any basic sequence into another basic sequence. Now if

$$p_n(x) = \sum_{i=0}^{n} a_i x^i$$

then

$$Tp_n(x) = T(\sum_{i=0}^{n} a_i x^i)$$

$$= \sum_{k=0}^{n} a_i Tx^i$$

$$= \sum_{k=0}^{n} a_i q_i(x)$$

$$= p_n(\underline{q}(x)) .$$

Thus $r_n(x)$ is a sequence of basic polynomials and by the Automorphism Theorem, it is the basic sequence for

$$R = TPT^{-1} = p(q(D)) , \qquad \text{Q. E. D.}$$

<u>Corollary</u>: If $p_n(x)$ is a sequence of basic polynomials then there exists a basic sequence $q_n(x)$ such that

$$p_n(\underline{q}(x)) = x^n .$$

We say that $q_n(x)$ is the <u>inverse sequence</u> of $p_n(x)$.

<u>Theorem 7. (Summation Formula).</u> Suppose $p_n(x)$ and $q_n(x)$ are the basic sequences for the delta operators P and Q respectively. If $q_n(x)$ is inverse to $p_n(x)$, then

$$p_n(x) = \sum_{k \geq 0} \frac{x^k}{k!} [Q^k x^n]_{x=0} .$$

The proof is similar to the preceding and is left to the reader.

We are now in a position to solve the problem stated in the Introduction: given basic sequences $p_n(x)$ and $q_n(x)$, with delta operators $P = p(D)$ and $Q = q(D)$, how are the coefficients c_{nk}

$$q_n(x) = \sum_{k\geq 0} c_{nk} p_k(x)$$

linking the $p_n(x)$ to the $q_n(x)$, the so-called <u>connection constants</u>, to be determined? The answer is dismayingly simple. Consider the polynomials

$$r_n(x) = \sum_{k\geq 0} c_{nk} x^k ,$$

and consider the umbral operator T defined by

$$Tx^n = p_n(x).$$

Then clearly

$$q_n(x) = Tr_n(x) = r_n(\underline{p}(x)) ,$$

so that $r_n(x)$ are of binomial type and $R = r(D)$ being their delta operator, we find $q(t) = r(p(t))$, or $r(t) = q(p^{-1}(t))$. Theorem 4 then provides explicit expressions for the $r_n(x)$. One couldn't expect a simpler answer.

As an example, consider the connection constants between $q_n(x) = x^n$ and $p_n(x) = (x)_n$. Here $q(t) = t$ and $p(t) = e^t - 1$. Thus, $r(t) = \log(1+t)$ and, as we shall see below, the polynomials $r_n(x)$ turn out to be the exponential polynomials $\varphi_n(x)$, discussed below.

As a second example, let $p_n(x) = (x)_n$ and $q_n(x) = x^{(n)}$. As easy computation shows that $r(t) = t/(t-1)$, whose basic polynomials are the Laguerre polynomials, also discussed below.

An instructive example the reader may work out for himself-thereby obtaining a number of classical and new identities, is to take $p_n(x) = x(x-na)^{n-1}$ and $q_n(x) = x(x-nb)^{n-1}$ for $a \neq b$. These examples could be multiplied ad infinitum, and a great number of combinatorial identities in the literature can be seen to fall into the simple pattern we have just outlined.

<u>Remark.</u> It can be shown that every automorphism

of the algebra Σ is of the form $S \to TST^{-1}$ for some umbral operator T, but this fact will not be needed, so we omit the proof.

8. The Exponential Polynomials.

The exponential polynomials, studied by Touchard and other authors, are a good testing ground for the theory developed so far. We shall see that their basic properties and the identities they satisfy are almost trivial consequences of the theory.

Consider the sequence of lower factorials $(x)_n$, which as we have seen is the basic sequence for the delta operator $\Delta = e^D - I$. In this case the inverse sequence is the sequence of basic polynomials for the operator $Q = \log(I+D)$. We denote these polynomials by $\varphi_n(x)$; these are the exponential polynomials. From the Corollary above we have umbrally

$$\underline{\varphi}(\underline{\varphi}-1)(\underline{\varphi}-2)\ldots(\underline{\varphi}-n+1) = x^n .$$

Further by the summation formula

$$\varphi_n(x) = \sum_{k\geq 0} \frac{x^k}{k!} [\Delta^k x^n]_{x=0}$$

$$= \sum_{k\geq 0} S(n,k) x^k ,$$

where $S(n, k)$ denote the Stirling numbers of the second kind. Now let us apply the Rodrigues formula to see what we get. Since $Q = \log(I+D)$, we have $Q' = (I+D)^{-1}$ and hence

$$\varphi_n(x) = x(Q')^{-1}\varphi_{n-1}(x)$$

$$= x(I+D)\varphi_{n-1}(x)$$

$$= x\varphi_{n-1}(x) + x\varphi'_{n-1}(x) ,$$

which is the recurrence formula for the exponential polynomials.

The next property of these exponential polynomials which we shall prove is expressed umbrally as

$$\varphi_{n+1}(x) = x(\underline{\varphi}+1)^n .$$

Let T be the umbral operator that takes $(x)_n$ into x^n, so that $Tx^n = \varphi_n(x)$. Hence

$$Tx(x-1)_{n-1} = xx^{n-1}$$

or changing n to $n+1$

$$Tx(x-1)_n = xx^n = xT(x)_n$$

can be rewritten as

$$TxE^{-1}(x)_n = xT(x)_n .$$

We can extend this by linearity to all polynomials $p(x)$ so that

$$TxE^{-1}p(x) = xTp(x) .$$

Replacing $p(x)$ by $p(x+1)$ we have

$$TxE^{-1}p(x+1) = xTp(x+1) .$$

Hence

$$Txp(x) = xTp(x+1).$$

Since $Tx^n = \varphi_n(x)$ then $Tp(x) = p(\underline{\varphi}(x))$ and it follows that

$$\begin{aligned}\underline{\varphi}(x)\,p(\underline{\varphi}(x)) &= Txp(x)\\ &= xTp(x+1\\ &= xp(\underline{\varphi}(x)+1).\end{aligned}$$

If we let $p(x) = x^n$ then

$$\varphi_{n+1}(x) = [\underline{\varphi}(x)]^{n+1} = \underline{\varphi}(x)[\underline{\varphi}(x)]^n$$
$$= x[\varphi(x)+1]^n$$

which is what we wanted to prove.

In a similar vein one can prove the remarkable Dobinsky-type formula:

$$\varphi_n(x) = e^{-x} \sum_{k \geq 0} \frac{x^k k^n}{k!},$$

which we shall leave as an exercise to the reader.

9. Laguerre Polynomials.

As a further example of the above theory, we shall develop some properties of the Laguerre polynomials. The Laguerre operator L is defined by

$$Lp(x) = -\int_0^\infty e^{-t} p'(x+t)dt .$$

It is a delta operator and as such has a sequence of basic polynomials which we shall call $L_n(x)$. By straightforward calculation, we find that the expansion of L in terms of D has coefficients

$$[Lx^n]_{x=0} = -n! , \quad n \geq 1$$
$$= 0, \quad n = 0$$

and hence we find that

$$L = \frac{D}{D-I} .$$

Hence from formula (3) of Theorem 4 we have

$$(*) \qquad L_n(x) = x(D-I)^n x^{n-1} .$$

Since for all polynomials $p(x)$ we also have

$$e^{x}De^{-x}p(x) = e^{x}(e^{-x}p'(x)-e^{-x}p(x))$$

$$= (D-I)p(x)$$

then $e^{x}De^{-x} = D-I$ and

$$e^{x}D^{n}e^{-x} = (D-I)^{n}.$$

Therefore we obtain the classical Rodrigues formula,

$$L_n(x) = xe^{x}D^{n}e^{-x}x^{n-1}.$$

From formula (*) we find by binomial expansion that

$$L_n(x) = \sum_{k=1}^{n} \frac{n!}{k!}\binom{n-1}{k-1}(-x)^k$$

where the coefficients

$$\frac{n!}{k!}\binom{n-1}{k-1}$$

are known as the (signless) Lah numbers. Our notation for the polynomials L_n corresponds to the notation in Bateman for the polynomials $L_n^{(-1)}$, that is

$$L_n^{(-1)}(x) = \frac{1}{n!}L_n(x).$$

We now come to the most important fact about the Laguerre polynomials. The indicator of L is

$$f(t) = \frac{t}{t-1}$$

and hence

$$f(f(t)) = \frac{\frac{t}{t-1}}{\frac{t}{t-1} - 1} = \frac{t}{t-t+1} = t$$

Thus, by the Automorphism Theorem we infer that the Laguerre polynomials are a self-inverse set. Thus, we have as an immediate consequence the beautiful identity

$$x^n = \sum_{k=1}^{n} \frac{n!}{k!} \binom{n-1}{k-1} (-1)^k L_k(x) = L_n(\underline{L}(x)).$$

Other identities concerning Laguerre polynomials stem from the fact that

$$L = \frac{D}{D-I}.$$

Since $L_n(x)$ are the basic polynomials of L we have

$$\frac{D}{D-I} L_n(x) = nL_{n-1}(x)$$

and hence the classical recurrence formula

$$L_n'(x) = n(D-I)L_{n-1}(x) .$$

In fact, if we expand $\frac{D}{D-I}$ into series form

$$\frac{D}{D-I} = -D \cdot \frac{I}{I-D} = -D(I+D+D^2+\ldots)$$

we can use this to get other known recurrence formulas.

10. A Glimpse of Combinatorics.

Although we intend to leave most of the combinatorial applications of the preceding theory to the second part of this work, we shall outline two typical results which we hope will orient the reader to applications to problems of enumeration, typical of the second part of this work.

Theorem 8. Let P be an invertible shift-invariant operator. Let $p_n(x)$ be a sequence of basic polynomials satisfying

$$[x^{-1}p_n(x)]_{x=0} = n[P^{-1}p_{n-1}(x)]_{x=0} ,$$

for all $n > 0$. Then $p_n(x)$ is the sequence of basic polynomials for the delta operator $Q = DP$.

Proof: Define the operator Q by $Q1 = 0$ and

$$Qp_n(x) = np_{n-1}(x)$$

and extending by linearity. Note that Q is shift-invariant. In terms of Q, the preceding identity can be rewritten in the form

$$[x^{-1}p_n(x)]_{x=0} = [P^{-1}Qp_n(x)]_{x=0} \, .$$

By linearity, this extends to an identity for all polynomials $p(x)$ - an argument we have often used in this work. Thus recalling that $[x^{-1}p(x)]_{x=0} = [Dp(x)]_{x=0}$ whenever $p(0)=0$, we have

$$[Dp(x)]_{x=0} = [P^{-1}Qp(x)]_{x=0}$$

for all polynomials $p(x)$, including those for which $p(0) \neq 0$. Setting $p(x) = q(x+a)$ we obtain, using the shift-invariance of P and Q,

$$\begin{aligned} Dq(a) &= [P^{-1}QE^a q(x)]_{x=0} \\ &= [E^a P^{-1} Qq(x)]_{x=0} \\ &= P^{-1}Qq(a) \end{aligned}$$

for all constants a. But this means that $D = P^{-1}Q$, or $Q = DP$, Q. E. D.

Corollary 1. Given any sequence of constants c_{n1}, $n = 1, 2, \ldots$, there exists a unique sequence of basic polynomials $p_n(x)$ such that $[x^{-1}p_n(x)]_{x=0} = c_{n1}$, that is,

$$p_n(x) = \sum_{k \geq 1} c_{nk} x^k, \quad n = 1, 2, \ldots \, .$$

Corollary 2. Let $g(x)$ be the indicator of Q in the above. Then $g = f^{-1}$, where $f(t) = \sum_{k \geq 0} c_{k,1} \frac{t^k}{k!}$.

Proof: From above

$$D = QP^{-1} = \sum_{k \geq 0} c_{k,1} \frac{Q^k}{k!} = f(Q)$$

and the result follows.

We now give some applications of the above theory.

Application 1. Let $t_{n,k}$ be the number of forests of rooted labeled trees (i.e., trees with a distinguished vertex) with n vertices and k components, then

$$A_n(x) = \sum_{k \geq 0} t_{n,k} x^k = x(x+n)^{n-1}.$$

Proof: Since $t_{n,1}$ is the number of rooted trees on n vertices, then $t_{n,1} = nA_{n-1}(1)$ since each such tree on n vertices may be obtained by mapping a forest on $n-1$ vertices onto a single new root vertex. The resulting root may be labeled in n ways, i.e., either by using a new symbol or by using one of the $n-1$ old symbols and replacing it by the new symbol. But this relation may be written

$$[x^{-1} A_n(x)]_{x=0} = n[EA_{n-1}(x)]_{x=0}$$

and hence the delta operator for A_n is DE^{-1} by Theorem 8. Thus the associated polynomials are the Abel polynomials $x(x+n)^{n-1}$.

Corollary (Cayley). The number of labeled trees on n vertices is n^{n-2}.

Proof: Since the number of rooted labeled trees is n^{n-1} the number of unrooted trees is n^{n-2} since each free tree can be labeled in n ways.

Application 2. Let S_n be a symmetric group on n symbols and let $c_{n,k}$ be the number of group elements which consist of precisely k cycles. If

$$C_n(x) = \sum_{k \geq 0} c_{n,k} x^k \quad \text{then} \quad C_n(x) = x^{(n)} .$$

<u>Proof:</u> We note that in this case $c_{n,1} = (n-1)!$ which is clearly the number of group elements consisting of just one cycle, and thus by Corollary 2 this is the required sequence.

<u>Functional Digraphs.</u> A digraph, D, (with loops permitted) on n symbols is a functional digraph if and only if it satisfies the following two postulates,

1) each component of D contains precisely one consistently directed circuit; and
2) each non-circuit edge is directed towards the circuit contained in its component.

An idempotent is a functional digraph all of whose components contain a distinguished vertex which meets every edge of that component.

<u>Application 3.</u> The polynomial $p_n(x) = \sum_{k \geq 0} \binom{n}{k} k^{n-k} x^k$ is of binomial type. Let $h_{n,k}$ be the number of idempotent on n symbols with precisely k components. Then $h_{n,k} = \binom{n}{k} k^{n-k}$ since the k distinguished vertices, V, may be chosen in $\binom{n}{k}$ ways and the remaining $n-k$ points may be directed into V in k^{n-k} ways. However, we may also view each idempotent as a structure generated by its components. It is interesting to note that the coefficients $h_{n,1} = n$ and the associated delta operator has indicator $f^{[-1]}(t)$ where $f(t) = te^t$. Thus these polynomials are the inverse sequence of the Abel polynomials. Several identities for them may be derived in much the same way as we related the exponential polynomials to the lower factorials in Section 6.

Anticipating some developments in the second part of this paper, we may state the following principle. In order to enumerate by a sequence $c_{n,1}$ a class of rooted trees, graded by the number of vertices, one forms the associated basic set, which will enumerate a class of reluctant functions, and then proceeds to apply Theorem 8 or a variant of it, which will reflect the "composition rule" of such class of trees. The connection constants between two polynomial sequences enumerating sequences of reluctant functions have a combinatorial interpretation in terms of "piecing together" one set of trees in terms of another. Thus our starting point in the second part of this work will be: given two families F_1 and F_2 of rooted labeled forests, in how many ways can be a member of F_2 be "pieced together" from members of F_1? The simplest case of this is Cayley's theorem above, where F_1 consists of a single edge and F_2 consists of all labeled rooted forests.

REFERENCES

1. Rota, G.-C., On the Foundations of Combinatorial Theory, I, Theory of Möbius functions, Zeit. für Wahr., 2 (1964), 340-368.

2. Crapo, H. H. and Rota, G.-C., On the Foundation of Combinatorial Theory, II, Combinatorial Geometries, M. I. T. Studies in Applied Mathematics (to appear).

3. Boas, R. P. and Buck, R. C., Polynomial Expansions of Entire Functions, Springer, New York, 1964.

4. Martin, W. T., On Expansions in Terms of a Certain General Class of Functions, Amer. J. Math., 58 (1936), 407-420.

5. Pringsheim, A., Zur Geschichte des Taylorschen Lehrsatzes, Bibliotheca Math., (3) 1 (1900), 433-479.

6. Steffensen, J. F., The Poweroid, an Extension of the Mathematical Notion of Power, Acta Math. 73 (1941), 333-366.

7. Touchard, J., Nombres Exponentiels et Nombres de Bernoulli, Canad. J. Math., 8 (1956), 305-320.

8. Riordan, J., Combinatorial Identities, Wiley, New York, 1968.

9. Al-Salaam, W.A., Operational Expressions for the Laguerre and other Polynomials, Duke Math J., 31 (1964), 127-142.

10. Carlitz, L. and Riordan, J., The Divided Central Differences of Zero, Canad. J. Math., 15 (1963), 94-100.

11. Hurwitz, A., Ueber Abel's Verallgemeinerung der binomischen Formel, Acta Math., 26 (1902), 199-203.

12. Lah, I., Eine neue Art von Zahlen, ihre Eigneschaften ... Mittelungsbl. Math. Stat., 7 (1955), 203-216.

13. Riordan, J., Inverse Relations and Combinatorial Identities, Amer. Math. Monthly, 71 (1964), 485-498.

14. Rota, G.-C., The Number of Partitions of a Set. Amer. Math. Monthly, 71 (1964), 498-504.

15. Moon, J.W., Counting Labelled Trees, A Survey of Methods and Results, A publication of the Department of Mathematics of the University of Alberta, 1969.

16. Pincherle, S. and Amaldi, Le Operazioni Distributive, Zanichelli, Bologna, 1900.

17. Pincherle, S., Opere Scelte, 2 vols., Cremonese, Rome, 1954.

18. Sylvester, J.J., Collected Mathematical Papers, Cambridge, 1904-1912, 4 vols.

19. Bell, E.T., Postulational Bases for the Umbral Calculus, Amer. J. Math., 62 (1940), 717-724.

20. Harary, F., A Seminar in Graph Theory, Holt, New York, 1967.

21. Harary, Marshall, Combinatorial Theory, Blaisdell, Waltham, 1967.

22. Harary, Frank, Graph Theory, Addison-Wesley, Reading, 1969.

23. Sheffer, I. M., Some properties of polynomials of type zero, Duke Math. J. 5 (1939) pp. 590-622.

24. Knuth, D., Another enumeration of trees, Can. J. Math. 20 (1968), 1077-1086.

25. Riordan, J. Forests of labeled trees, J. C. T. 5 (1968) 90-103.

University of Colorado
Florida Atlantic University
Massachusetts Institute of Technology
University of Waterloo

A Composition Theorem for the Enumeration of Certain Subsets of the Symmetric Semigroup

BERNARD HARRIS AND LOWELL SCHOENFELD

1. Introduction.

For $n \geq 1$, let X_n be the set of integers $1, 2, \ldots, n$, and let T_n be the set of all mappings of X_n into itself. If $\alpha, \beta \in T_n$ then their product $\alpha\beta \in T_n$ is defined by $(\alpha\beta)(x) = \alpha(\beta(x))$ for all $x \in X_n$. With the product defined in this way, T_n is a semigroup with an identity - the so-called symmetric semigroup on n letters. T_n contains the set S_n of all mappings of X_n onto itself; S_n is, of course, the symmetric group whose elements permute the integers $1, 2, \ldots, n$.

In this paper we prove a very general result concerning exponential generating functions which enumerate the number of elements in certain subsets of T_n. This result may be described very roughly as follows. We suppose that $W_{\ell 1}, W_{\ell 2}, \ldots$ are certain given sets of permutations acting on ℓ letters and that each set has w_ℓ elements. Let

$$\Phi_W(z) = \sum_{\ell=1}^{\infty} \frac{1}{\ell!} w_\ell z^\ell \quad . \tag{1.1}$$

Associated with the given sets $W_{\ell r}$ is the quantity $V_{W;n;h}$ which is the number of $\alpha \in T_n$ satisfying certain conditions which will be made more specific later. These numbers have an exponential generating function

$$\Psi_{W;h}(z) = \sum_{n=1}^{\infty} \frac{1}{n!} V_{W;n;h} z^n \; . \tag{1.2}$$

We let $G_0(z) = z$ and $G_h(z) = z \exp G_{h-1}(z)$ for $h \geq 1$. One of the principal results of this paper is the composition formula

$$\Psi_{W;h}(z) = \Phi_W(G_h(z)) . \tag{1.3}$$

The results we obtain also have interpretations in terms of functional digraphs. In this case, $V_{W;n;h}$ is the number of such digraphs with n vertices and height not exceeding h whose cyclical vertices specify a digraph lying in the prescribed sets $W_{\ell r}$ for some $\ell \geq 1$ and some $r \geq 1$. The simplest case is that in which $W_{\ell r}$ consists of the identity map on the ℓ letters in question; hence all the cyclical elements have order 1. Interpreting the cycles as roots of trees, the α being counted are essentially forests of rooted labeled trees of height not exceeding h.

We may therefore think of the general case as being one in which we generalize the notion of root so that directed cycles are permitted instead of cycles of length 1. Now $V_{W;n;h}$ counts forests of generalized rooted labeled trees with n vertices and height not exceeding h and which are such that the generalized roots are specified by the sets $W_{\ell r}$.

Aside from various papers dealing with trees, the most closely related previous works are those of Katz [13], Harris [6], and Riordan [19].

In addition to proving various generalizations of (1.3), we apply our results to various examples and thereby obtain both new and old results.

2. Preliminaries.

In this section, we define the numbers $V_{W;n;h}$ together with other quantities and give interpretations of these in terms of graph theory.

Let α be a fixed mapping of X_n into itself, and for each $x \in X_n$ let us define

$$x_0 = x, \quad x_1 = \alpha(x), \quad x_2 = \alpha(x_1) = \alpha^2(x), \quad \ldots ;$$

in general, $x_{m+1} = \alpha(x_m)$ so that $x_{m+1} = \alpha^m(x_1) = \alpha^{m+1}(x)$

for all $m \geq 0$. We say that two elements $x, y \in X_n$ are <u>equivalent with respect to α</u> if there exist non-negative integers r and s such that $x_r = y_s$. It is easily verified that this is, in fact, an equivalence relation in X_n; the equivalence class containing x we designate by $K_\alpha(x)$ and call it the <u>component of x with respect to α</u>. If there is only one such equivalence class then α is said to be <u>connected.</u> Since $K_\alpha(x)$ is a finite set, we necessarily have the existence of two distinct non-negative integers r and s not exceeding n such that $x_r = x_s$; if $s > r$ and $y = x_r$, then $y \in K_\alpha(x)$ and $y_{s-r} = x_s = x_r = y$. Putting $t = s-r$, we have $1 \leq t \leq n$ and $y_t = y$. Consequently, there is a smallest $q \geq 1$ such that $y_q = y$. In such a case, we say that y is <u>cyclic with respect to α</u> and that y has <u>order q under α</u>. Thus, each $K_\alpha(x)$ contains at least one cyclic element.

Let y be cyclic with respect to α and have order q. Then we define $C_\alpha(y) = \{y_m \mid m \geq 0\}$ and call this the <u>α-cycle containing y</u>; the number of elements in $C_\alpha(y)$ is called its <u>length</u>. Clearly, the length of $C_\alpha(y)$ is q and $C_\alpha(y) \subset K_\alpha(y)$. Standard arguments show that $y_m = y$ if and only if $q \mid m$; hence, also, $y_m = y_k$ if and only if $m \equiv k \pmod q$. As a result, each y_m has order q; also, if $w = y_m$, then $C_\alpha(y) = C_\alpha(w)$. Further, we find it convenient to define C_α as the set of all elements which are cyclic with respect to α.

Now suppose y and z are two cyclic elements in $K_\alpha(x)$. Then for some r and s we have $y_r = z_s \equiv w$ say. Hence $C_\alpha(y) = C_\alpha(w) = C_\alpha(z)$. Thus each component $K_\alpha(x)$ contains exactly one α-cycle; this cycle we designate by $D_\alpha(x)$. If x is cyclic with respect to α, then so is x_m for each m and $D_\alpha(x) = C_\alpha(x) = C_\alpha(x_m)$. Further, $C_\alpha = \bigcup_{x \in X_n} D_\alpha(x)$.

As we have seen, to each $x \in X_n$ there corresponds some integer $r \geq 0$ such that $y = x_r$ is cyclic with respect to α. The smallest such r is called the <u>α-height of x</u> which we designate by $h_\alpha(x)$. We easily see that $x_0, x_1, \ldots, x_{h_\alpha(x)}$ are all distinct and $0 \leq h_\alpha(x) \leq n-1$. Also, $h_\alpha(x)=0$

if and only if x is cyclic with respect to α. Finally, we call $\max_{x \in X_n} h_\alpha(x)$ the <u>height of α</u>.

With each $\alpha \in T_n$ we associate the mapping α^*, the restriction of α to C_α, defined by $\alpha^*(y) = \alpha(y)$ if $y \in C_\alpha$. Clearly, $\alpha^* = \alpha$ if and only if every $x \in X_n$ is cyclic with respect to α; also $\alpha^*(y) \in C_\alpha$. Moreover, α^* permutes the elements of C_α. To see this, suppose that $x, y \in C_\alpha$ are such that $\alpha^*(x) = \alpha^*(y)$. Then, on letting q be the least common multiple of the orders of x and y, we have

$$x = \alpha^q(x) = \alpha^{q-1}(\alpha(x)) = \alpha^{q-1}(\alpha^*(x)) = \alpha^{q-1}(\alpha^*(y))$$

$$= \alpha^{q-1}(\alpha(y)) = \alpha^q(y) = y .$$

So, $\alpha^*(x) = \alpha^*(y)$ implies $x = y$; as C_α is a finite set, α^* is a permutation. Dénes [3, 4, 5] calls α^* the <u>main permutation of α</u>. However α^* need not be in S_n since it is defined only on C_α which is usually a proper subset of X_n; indeed, α^* need not be in any of $S_1, S_2, \ldots, S_n$. Here, and later, we use the term <u>permutation</u> to mean a one-to-one mapping of some finite set onto itself.

Roughly speaking, the quantity $V_{W;n;h}$ will count all $\alpha \in T_n$ of height not exceeding h such that the main permutation α^* satisfies certain restrictions indicated by W. In order to formulate these restrictions, it is convenient to list all subsets of the set N of positive integers.

For a fixed $\ell \geq 1$, we begin by enumerating those subsets having exactly ℓ elements. As $X_1, X_2, \ldots, X_{\ell-1}$ contain no such subsets, we start with X_ℓ which has exactly one such subset, namely itself, and we define $X_{\ell 1} = X_\ell$. Next, $X_{\ell+1}$ has $\nu_1 = \binom{\ell+1}{\ell}$ such subsets of which $\nu_1 - 1$ are new subsets; we order these $\nu_1 - 1$ in some way (which way being immaterial for our purposes) and call them $X_{\ell 2}, X_{\ell 3}, \ldots, X_{\ell \nu_1}$. Further, $X_{\ell+2}$ has $\nu_2 = \binom{\ell+2}{\ell}$ distinct subsets with ℓ elements and $\nu_2 - \nu_1 = \binom{\ell+1}{\ell-1}$ of these subsets are new ones which we call $X_{\ell, \nu_1+1}, \ldots, X_{\ell, \nu_2}$. We continue in this way thereby obtaining, for $\ell \geq 1$, the sequence of distinct sets

(2.1) $$X_{\ell 1},\ X_{\ell 2},\ X_{\ell 3},\ \ldots$$

consisting of all subsets of N having exactly ℓ elements. Further, $X_{\ell 1} = X_\ell$.

The sets in (2.1) will form the ℓ-th row of a matrix X . It is also convenient to introduce a 0-th row consisting of the sets $X_{01}, X_{02}, \ldots$ each of which is the empty set ϕ ; i.e. $X_{0r} = \phi$ for all $r \geq 1$. Thus we define the matrix

(2.2) $$X = \begin{pmatrix} X_{01} & X_{02} & X_{03} & X_{04} & \cdots \\ X_{11} & X_{12} & X_{13} & X_{14} & \cdots \\ X_{21} & X_{22} & X_{23} & X_{24} & \cdots \\ \vdots & \vdots & \vdots & \vdots & \end{pmatrix}$$

which lists all the subsets $X_{\ell r}$ of N; and if we disregard $X_{02}, X_{03}, \ldots$ then all the subsets of N appear exactly once. It is clear, for example, that X_{1r} consists of the single number r .

We now introduce a matrix

(2.3) $$W = \begin{pmatrix} W_{01} & W_{02} & W_{03} & W_{04} & \cdots \\ W_{11} & W_{12} & W_{13} & W_{14} & \cdots \\ W_{21} & W_{22} & W_{23} & W_{24} & \cdots \\ \vdots & \vdots & \vdots & \vdots & \end{pmatrix}$$

of sets having $W_{0r} = \phi$ for all $r \geq 1$. Of the sets $W_{\ell r}$ for $\ell \geq 1$ we only require that its elements be permutations of $X_{\ell r}$; any of these $W_{\ell r}$ may be ϕ . As $X_{\ell r}$ has exactly ℓ elements selected from N , the elements of $W_{\ell r}$ are certain permutations on ℓ elements selected from N. Since $X_{\ell 1} = X_\ell$, we see that the elements of $W_{\ell 1}$ are permutations of X_ℓ; i.e. $W_{\ell 1} \subset S_\ell$. Moreover, if $\ell \geq 1$ and $r \geq 2$, then $X_{\ell r}$ contains some $x \notin X_\ell$; consequently, $W_{\ell r} \cap T_\ell = \emptyset$ if $\ell \geq 1$ and $r \geq 2$. We also note that all the sets $W_{\ell r}$ listed by W are disjoint.

We let $w_{\ell r}$ be the number of elements in $W_{\ell r}$ so that $w_{0r} = 0$ and $0 \leq w_{\ell r} \leq \ell!$ for all $\ell \geq 0$ and $r \geq 1$. We also define $w_\ell = w_{\ell 1}$ for $\ell \geq 0$. Then $0 \leq w_\ell \leq \ell!$ so that $\Phi_W(z)$, defined by (1.1), is holomorphic for $|z| < 1$.

We can now define $V_{W;n;h}$. Corresponding to a given matrix W, to an integer $n \geq 1$, and to an integer $h \geq 0$ is the number $V_{W;n;h}$ of all $\alpha \in T_n$ such that:

(A') $\alpha^* \in W_{\ell r}$ for some $\ell \geq 0$ and some $r \geq 1$

(B) α has height at most h.

Clearly $0 \leq V_{W;n;h} \leq n^n$; and if $h \geq n-1$, then $V_{W;n;h} = V_{W;n;n-1}$. Also, $\Psi_{W;h}(z)$, defined by (1.2), is holomorphic for $|z| < 1/e$.

It is clear that we can not expect to be able to say very much about $V_{W;n;h}$ unless we place restrictions on the sets $W_{\ell r}$ over which α^* is permitted to vary. For this reason, we will mostly deal with matrices W such that for each $\ell \geq 0$, $w_{\ell r} = w_\ell$ for all $r \geq 1$. Such a W is said to have the <u>cardinality property</u>. Thus, W has the cardinality property if and only if for every $\ell \geq 1$ all the prescribed sets $W_{\ell r}$, $r \geq 1$, of permutations on ℓ letters have exactly the same number w_ℓ of elements. Our main object is to prove that (1.3) holds if W has the cardinality property and $|z| \leq 1/4$.

It is convenient to reformulate these notions in terms of graph theory. We do this by setting up a one-to-one correspondence between T_n and the set of all functional digraphs with n vertices; here, a functional digraph means a directed graph with exactly one edge emanating from each vertex. Given $\alpha \in T_n$, we associate with it the functional digraph $G(\alpha)$ by drawing a directed edge from x to $\alpha(x)$ for all those $x \in X_n$ having $\alpha(x) \neq x$; if $\alpha(x) = x$, then we draw a directed loop through x. The resulting figure is $G(\alpha)$. Conversely, it is apparent that each functional digraph **G** with n vertices determines a unique $\alpha \in T_n$ such that $G(\alpha) = G$.

Because of this correspondence between the mappings of X_n into X_n and the functional digraphs whose vertices are in X_n, we can identify α and $G(\alpha)$; thus, where convenient, we will speak of the graph α and the mapping $G(\alpha)$. T_n will also be referred to as the set of all functional digraphs with vertices in X_n. This usage is consistent with the fact that the set $K_\alpha(x)$ is the graph-theoretic component of $G(\alpha)$ which contains the vertex x. And $C_\alpha(y)$ is just the graph-theoretic cycle of $G(\alpha)$ which contains the cyclic vertex y. Also, the height of α is just the height of the graph $G(\alpha)$.

Further, the main permutation α^* corresponds to the subgraph obtained from $G(\alpha)$ by deleting the non-cyclic vertices and the edges emanating from them. In our present terminology, the elements of the sets $W_{\ell r}$ are functional digraphs having ℓ vertices selected from N and having the additional property that all ℓ vertices are cyclical. The cardinality property requires that each of $W_{\ell 1}, W_{\ell 2}, \ldots$ contain the same number w_ℓ of graphs. As a result, we see that $V_{W;n;h}$ is just the number of functional digraphs α with vertices in X_n such that:

(A') $\qquad \alpha^* \in W_{\ell r}$ for some $\ell \geq 0$ and some $r \geq 1$

(B) $\qquad \alpha$ has height at most h.

The sets $W_{\ell r}$ are to be thought of as prescribed ones so that (A') imposes the restraint on the graphs α being enumerated by $V_{W;n;h}$ that the main permutation α^* should be in the union $\bigcup_{\ell=0}^{\infty} \bigcup_{r=1}^{\infty} W_{\ell r}$ of all sets in the matrix W.

In the case in which $W_{\ell r}$ consists of the identity mapping on $X_{\ell r}$, we have the situation in which every cyclical element has order 1. If the cyclicical elements of $G(\alpha)$ are taken to be roots, then $G(\alpha)$ becomes a forest of rooted labeled trees of height at most h.

Consequently, $V_{W;n;h}$ is just the number of forests of rooted labeled trees witn n vertices and height at most h. In the general case for W, each cyclical element will no longer correspond to a root but we may think of it as being part of some (directed) root structure specified by W. Therefore, we may think of $V_{W;n;h}$ as counting the total number of forests of labeled trees with n vertices and height not exceeding n with the root structure given by W.

In addition to $V_{W;n;h}$, there are a number of other quantities of interest. For example, if $n \geq 1$, $h \geq 0$, $H \geq 0$ and $\ell \geq 0$ then we define $Z_{W;n;h,H,\ell}$ as the number of $\alpha \in T_n$ such that:

(A) $\alpha^* \in W_{\ell r}$ for some $r \geq 1$

(B) α has height at most h

(C) α has exactly H vertices with height h.

In order to simplify the notation, we put $Z_{h,H,\ell} = Z_{W;n;h,H,\ell}$ where convenient. Likewise, we write $V_h = V_{W;n;h}$. Of course,

$$V_h = \sum_{H=0}^{n} \sum_{\ell=0}^{n} Z_{h,H,\ell} .$$

Clearly

(2.4) $$Z_{h,H,\ell} = 0$$

if any of the following hold:

(2.5) $$\ell = 0$$

(2.6) $$\ell \geq n+1$$

(2.7) $$H = 0 = h$$

(2.8) $$H \geq 1 \text{ and } h \geq n$$

(2.9) $H \geq n+1$

(2.10) $h = 1$ and $H \neq n-\ell$

(2.11) $H \geq n+1-\ell$ and $h \geq 1$.

For example, if (2.11) holds but (2.4) is false, then there exist $\alpha \in T_n$ having at least $n+1-\ell$ vertices of positive height and, by (A), just ℓ vertices of height 0 ; thus α has at least $n+1$ distinct vertices, and this contradicts $\alpha \in T_n$.

Next, let $V_{h,H} = V_{W;n;h,H}$ be the number of $\alpha \in T_n$ such that:

(A') $\alpha^* \in W_{\ell r}$ for some $\ell \geq 0$ and some $r \geq 1$

(B) α has height at most h

(C) α has exactly H vertices with height h.

Then we have

(2.12) $V_{h,H} = 0$ if any of (2.7)-(2.9) hold

(2.13) $$V_{h,H} = \sum_{\ell=0}^{n} Z_{h,H,\ell}\ .$$

Also, our previous definition of $V_{W;n;h} = V_h$ shows that

(2.14) $$V_h = \sum_{H=0}^{n} V_{h,H}$$

(2.15) $V_h = V_{n-1}$ if $h \geq n-1$.

We also define $Y_{h,\ell} = Y_{W;n;h,\ell}$ as the number of $\alpha \in T_n$ such that:

(A) $\alpha^* \in W_{\ell r}$ for some $r \geq 1$

(B) α has height at most h.

Then:

(2.16) $Y_{h,\ell} = 0$ if either (2.5) or (2.6) holds

(2.17) $Y_{h,\ell} = Y_{n-1,\ell}$ if $h \geq n-1$

(2.18) $$Y_{h,\ell} = \sum_{H=0}^{n} Z_{h,H,\ell}$$

(2.19) $Y_{1,\ell} = Z_{1,n-\ell,\ell}$ if $0 \leq \ell \leq n$

(2.20) $$V_h = \sum_{\ell=0}^{n} Y_{h,\ell} .$$

Further, we define $Y_\ell = Y_{W;n;\ell}$ as the number of $\alpha \in T_n$ such that:

(A) $\alpha^* \in W_{\ell r}$ for some $r \geq 1$.

Then

(2.21) $Y_\ell = 0$ if either (2.5) or (2.6) holds

(2.22) $Y_\ell = Y_{h,\ell}$ if $h \geq n-1$.

Finally, we let $Z = Z_{W;n}$ be the number of $\alpha \in T_n$ such that:

(A') $\alpha^* \in W_{\ell r}$ for some $\ell \geq 0$ and some $r \geq 1$.

Hence,

(2.23) $$Z = \sum_{\ell=0}^{n} Y_\ell$$

(2.24) $Z = V_h$ if $h \geq n-1$.

We may also note that

(2.25) $$Y_{W;n;0} = w_n .$$

For the left side is the number of $\alpha \in T_n$ such that $\alpha^* \in W_{nr}$ for some $r \geq 1$. Hence the left side is the number of $\alpha \in T_n$ such that $\alpha \in W_{nr}$. Now $W_{nr} \cap T_n = \phi$ if $r \geq 2$. Hence $Y_{W;n;n}$ is just the number of $\alpha \in W_{n1}$; the last set has $w_{n1} = w_n$ elements by definition so that (2.25) holds. Similarly,

(2.26) $$V_{W;n;0} = w_n .$$

3. The Composition Theorem.

This result, of which (1.3) is a special case, employs an enumeration by height, which is facilitated by the introduction of the quantity $N_{W;n}(\ell, \ell_1, \ldots, \ell_h)$ which we simplify to $N(\ell, \ell_1, \ldots, \ell_h)$. If $n \geq 1$, $h \geq 0$, and $\ell, \ell_1, \ldots, \ell_h$ are non-negative integers such that

(3.1) $$\ell + \ell_1 + \ldots + \ell_h = n ,$$

then $N(\ell, \ell_1, \ldots, \ell_h)$ is the number of $\alpha \in T_n$ such that:

(A) $\alpha^* \in W_{\ell r}$ for some $r \geq 1$

(C') For $m = 1, \ldots, h$ the functional digraph α has exactly ℓ_m vertices with height m.

If $h = 0$, then (C') is a vacuous condition so that (3.1) and (2.25) yield

(3.2) $$N_{W;n}(n) = Y_{W;n;n} = w_n .$$

For $h = 1$, we obtain

(3.3) $$N(\ell, n-\ell) = Y_{1,\ell} = Z_{1,n-\ell,\ell} \quad \text{if } 0 \leq \ell \leq n$$

as a consequence of (2.19). Also, since $W_{0r} = \phi$ for all $r \geq 1$, we have

(3.4) $$N(0, \ell_1, \ldots, \ell_h) = 0 \quad \text{if } h \geq 1 .$$

It may also be noted that if $\alpha \in T_n$ satisfies (A) or (A'), then the domain of α^* consists of exactly ℓ elements $x \in X_n$ and these are cyclic and hence have an α-height of 0. Moreover, there are exactly $n-\ell$ elements in X_n having a positive α-height. If, in addition, α satisfies (C') and (3.1) or if α satisfies (B), then the $n-\ell$ elements have heights between 1 and h.

Theorem 1. If $h \geq 2$, $H \geq 0$ and $\ell \geq 0$, then

$$(3.5) \qquad Z_{h, H, \ell} = \sum_{\substack{\ell_1+\dots+\ell_{h-1} = n-\ell-H \\ \ell_1, \dots, \ell_{h-1} \geq 0}} N(\ell, \ell_1, \dots, \ell_{h-1}, H).$$

And if $h \geq 1$, $\ell_0 \geq 0$ and W has the cardinality property,

$$(3.6) \qquad N_{W;n}(\ell_0, \dots, \ell_h) = \frac{n!}{\ell_0! \dots \ell_h!} \ell_0^{\ell_1} \dots \ell_{h-1}^{\ell_h} w_{\ell_0}$$

$$= \binom{n}{\ell_h} \ell_{h-1}^{\ell_h} N_{W;n-\ell_h}(\ell_0, \dots, \ell_{h-1}).$$

Proof. If α is counted by $Z_{h, H, \ell}$, then α satisfies (A) so that it has ℓ vertices with height 0; as (C) holds, α has H vertices of height $h \geq 2$. Hence α has $n-\ell-H$ vertices with heights between 1 and $h-1$. If we let ℓ_m be the number of vertices with height m for $1 \leq m \leq h-1$, then $\ell_1+\dots+\ell_{h-1} = n-\ell-H$ and $\ell_m \geq 0$. Consequently, α will be counted by the sum on the right of (3.5); in fact, α will be counted exactly once by the summands in (3.5). Conversely, every α counted by these summands satisfies (A), (B), (C) and hence is counted by $Z_{h, H, \ell}$. This proves (3.5).

Now suppose that W has the cardinality property and that $\ell_0+\dots+\ell_h = n$ where each $\ell_m \geq 0$ and $h \geq 0$. We first note that the number, ν, of ways of distributing the n elements of X_n into $h+1$ disjoint sets $R_0, \dots, R_h$ with $\ell_0, \dots, \ell_h$ elements, respectively, is given by

$$\nu = \frac{n!}{\ell_0!\ell_1!\ldots\ell_h!}.$$

With the h+1 sets R_m having ℓ_m elements now fixed, we determine the number λ of $\alpha \in T_n$ such that the elements of $R_h, \ldots, R_1$ have α-heights of $h, \ldots, 1$ and such that $\alpha^* \in W_{\ell_0, r}$ for some $r \geq 1$. The last condition is equivalent to requiring that α have exactly ℓ_0 vertices of height 0 and these vertices must therefore constitute the set $R_0 = X_{\ell_0, r}$ for some $r \geq 1$. Inasmuch as the α-height of $\alpha(x)$ is $h_\alpha(x)-1$ whenever x is not cyclical with respect to α, it is clear that the α we are attempting to count must map R_m into R_{m-1} for $m = h, \ldots, 2$ and α must also map R_1 into the remaining set R_0 whose elements have height 0. Now there are $\ell_{h-1}^{\ell_h}$ ways of mapping the ℓ_h elements of R_h into the ℓ_{h-1} elements of R_{h-1}; likewise, R_{h-1} can be mapped into R_{h-2} in $\ell_{h-2}^{\ell_{h-1}}$ ways; ...; and R_1 can be mapped into R_0 in $\ell_0^{\ell_1}$ ways. Finally, because $R_0 = X_{\ell_0, r}$ has all its elements of height 0, the number of mappings of R_0 onto R_0 which satisfy (A) is exactly w_{ℓ_0} as a result of the cardinality property. Consequently, the number λ of mappings is given by

$$\lambda = \ell_{h-1}^{\ell_h}\ell_{h-2}^{\ell_{h-1}}\cdots\ell_0^{\ell_1}w_{\ell_0}.$$

Inasmuch as $N(\ell_0, \ldots, \ell_h) = \nu\lambda$ we see that the first equation in (3.6) holds for $h \geq 0$. Hence, the last part of (3.6) holds for $h \geq 1$.

<u>Corollary 1.</u> If $\ell \geq 0$, then

(3.7) $$Y_{h,\ell} = \sum_{\substack{\ell_1+\ldots+\ell_h=n-\ell \\ \ell_1,\ldots,\ell_h \geq 0}} N(\ell, \ell_1, \ldots, \ell_h) \quad \text{if } h \geq 1$$

(3.8) $$V_h = \sum_{\substack{\ell_0+\ldots+\ell_h=n \\ \ell_0,\ldots,\ell_h \geq 0}} N(\ell_0, \ldots, \ell_h) \quad \text{if } h \geq 0.$$

Proof. As (3.7) clearly holds if $\ell \geq n+1$, we assume that $\ell \leq n$. If $h=1$, (3.7) follows from (3.3). For $h \geq 2$, (3.7) follows from (2.18) and (3.5). Then (3.8) is a consequence of (2.20) and (3.7) if $h \geq 1$; for $h=0$, it suffices to use (3.2) and (2.26).

Corollary 2. If $\ell \geq 0$ and W has the cardinality property, then

$$Y_\ell = \binom{n}{\ell} \ell\, w_\ell\, n^{n-\ell-1}. \tag{3.9}$$

Proof. If $\ell = 0$ or $\ell \geq n+1$, then (3.9) holds since both sides are 0 by (2.21). If $\ell = n$, then both sides of (3.9) are w_n by (2.25). It therefore suffices to consider $n \geq 2$ and $1 \leq \ell \leq n-1$. By (2.22), (3.7), and (3.6)

$$Y_\ell = Y_{n-1,\ell} = \sum_{\substack{\ell_1+\dots+\ell_{n-1}=n-\ell \\ \ell_1,\dots,\ell_{n-1}\geq 0}} N(\ell, \ell_1, \dots, \ell_{n-1})$$

$$= \frac{n!}{\ell!} w_\ell \sum_{\substack{\ell_1+\dots+\ell_{n-1}=n-\ell \\ \ell_1,\dots,\ell_{n-1}\geq 0}} S(\ell_1, \dots, \ell_{n-1}), \tag{3.10}$$

where

$$S(\ell_1, \dots, \ell_{n-1}) = \frac{\ell^{\ell_1} \ell_1^{\ell_2} \cdots \ell_{n-2}^{\ell_{n-1}}}{\ell_1!\,\ell_2! \cdots \ell_{n-1}!}.$$

If not all of $\ell_1, \dots, \ell_{n-2}$ are positive, there is a smallest $m \geq 0$ such that $\ell_{m+1} = 0$. In this case, $S(\ell_1, \dots, \ell_{n-1}) = 0$ unless $0 = \ell_{m+2} = \dots = \ell_{n-1}$. As summands having the value 0 can be discarded from (3.10), we need only consider those for which $\ell_1 \geq 1, \dots, \ell_m \geq 1$ and $0 = \ell_{m+1} = \dots = \ell_{n-1}$; if these relations hold, then

$$S(\ell_1, \dots, \ell_{n-1}) = \frac{\ell^{\ell_1} \ell_1^{\ell_2} \cdots \ell_{m-1}^{\ell_m}}{\ell_1!\,\ell_2! \cdots \ell_m!}$$

and $\ell_1 + \dots + \ell_m = n-\ell \geq 1$ so that, necessarily, $m \geq 1$ and $m \leq n-\ell$. Hence (3.10) shows that Y_ℓ is equal to

$$\binom{n}{\ell} w_\ell \sum_{m=1}^{n-\ell} \sum_{\substack{\ell_1+\dots+\ell_m=n-\ell \\ \ell_1,\dots,\ell_m \geq 1}} \frac{(n-\ell)!}{\ell_1!\,\ell_2!\cdots\ell_m!}\, \ell^{\ell_1} \ell_1^{\ell_2} \cdots \ell_{m-1}^{\ell_m}.$$

The corollary now follows on taking $q = n-\ell \geq 1$ in the following result of Katz [13] whose proof contained flaws:

Lemma 1. For all complex z and integral $q \geq 1$

$$(3.11) \qquad z(z+q)^{q-1} = \sum_{m=1}^{q} \sum_{\substack{\ell_1+\dots+\ell_m=q \\ \ell_1,\dots,\ell_m \geq 1}} \frac{q!}{\ell_1!\cdots\ell_m!}\, z^{\ell_1} \ell_1^{\ell_2} \cdots \ell_{m-1}^{\ell_m}.$$

Proof. We give a modified form of Katz' proof. Clearly the result holds for $q = 1$. We proceed by induction assuming it holds for all integers in $[1, q-1]$ where $q \geq 2$. Now

$$z(z+q)^{q-1} = z \sum_{\ell_1=1}^{q} \binom{q-1}{\ell_1-1} z^{\ell_1-1} q^{q-\ell_1}$$

$$= z^q + \sum_{\ell_1=1}^{q-1} \binom{q}{\ell_1} z^{\ell_1} \cdot \ell_1 \{\ell_1 + (q-\ell_1)\}^{q-\ell_1-1}.$$

As $1 \leq q-\ell_1 \leq q-1$, we can apply the induction hypothesis to get

$$z(z+q)^{q-1}$$

$$= z^q + \sum_{\ell_1=1}^{q-1} \binom{q}{\ell_1} z^{\ell_1} \sum_{m=1}^{q-\ell_1} \sum_{\substack{\ell_2+\dots+\ell_{m+1}=q-\ell_1 \\ \ell_2,\dots,\ell_{m+1}\geq 1}} \frac{(q-\ell_1)!}{\ell_2!\dots\ell_{m+1}!} \ell_1^{\ell_2}\dots\ell_m^{\ell_{m+1}}$$

$$= z^q + \sum_{m=1}^{q-1} \sum_{\ell_1=1}^{q-m} \sum_{\substack{\ell_2+\dots+\ell_{m+1}=q-\ell_1 \\ \ell_2,\dots,\ell_{m+1}\geq 1}} \frac{q!}{\ell_1!\ell_2!\dots\ell_{m+1}!} z^{\ell_1} \ell_1^{\ell_2}\dots\ell_m^{\ell_{m+1}}$$

$$= z^q + \sum_{M=2}^{q} \sum_{\substack{\ell_1+\dots+\ell_M=q \\ \ell_1,\dots\ell_M\geq 1}} \frac{q!}{\ell_1!\dots\ell_M!} z^{\ell_1} \ell_1^{\ell_2}\dots\ell_{M-1}^{\ell_M} .$$

As z^q corresponds to $M = 1$, we obtain (3.11) and the induction is complete.

We now introduce the homogenous polynomial in $t_0, \dots, t_h$ of degree n given by

(3.12) $$V_{W;n;h}(t_0, \dots, t_h) = \sum_{\substack{\ell_0+\dots+\ell_h=n \\ \ell_0,\dots,\ell_h\geq 0}} N_{W;n}(\ell_0, \dots, \ell_h) t_0^{\ell_0} \dots t_h^{\ell_h}$$

if $h \geq 0$ and $n \geq 1$; we abbreviate this to $V_h(t_0, \dots, t_h)$. By (3.8)

(3.13) $$V_h(1, \dots, 1) = V_h .$$

If $\zeta = \max \{|zt_0|, \dots, |zt_h|\}$, then

$$|z^n V_h(t_0, \ldots, t_h)|$$

$$= \left| \sum_{\substack{\ell_0 + \ldots + \ell_h = n \\ \ell_0, \ldots, \ell_h \geq 0}} N(\ell_0, \ldots, \ell_h)(zt_0)^{\ell_0} \ldots (zt_h)^{\ell_h} \right|$$

$$\leq \sum_{\substack{\ell_0 + \ldots + \ell_h = n \\ \ell_0, \ldots, \ell_h \geq 0}} N(\ell_0, \ldots, \ell_h) \zeta^{\ell_0} \ldots \zeta^{\ell_h}$$

$$= \zeta^n V_h \leq \zeta^n n^n .$$

Consequently, if we set

$$\Psi_{W;h}(z;t_0, \ldots, t_h) = \sum_{n=1}^{\infty} \frac{1}{n!} V_{W;n;h}(t_0, \ldots, t_h) z^n , \tag{3.14}$$

then this generating function has an absolutely convergent expansion if $\zeta < 1/e$. Also, (3.12) gives

$$\Psi_{W;h}(z;t_0, \ldots, t_h) \tag{3.15}$$

$$= \sum_{n=1}^{\infty} \frac{1}{n!} z^n \sum_{\substack{\ell_0 + \ldots + \ell_h = n \\ \ell_0, \ldots, \ell_h \geq 0}} N_{W;n}(\ell_0, \ldots, \ell_h) t_0^{\ell_0} \ldots t_h^{\ell_h}$$

$$\Psi_{W;h}(z;1, \ldots, 1) = \Psi_{W;h}(z) \tag{3.16}$$

as a result of (3.13) and (1.2). The following result is a further step towards the composition theorem.

<u>Lemma 2.</u> Let $h \geq 1$, $\max\{zt_0|, \ldots, |zt_h|\} < 1/e$ and $|zt_{h-1} e^{zt_h}| < 1/e$. If W has the cardinality property, then

$$(3.17)\quad \Psi_{W;h}(z;t_0,\ldots,t_h) = \Psi_{W;h-1}(z;t_0,\ldots,t_{h-2},t_{h-1}e^{zt_h})$$

Proof. By (3.12)

$$V_{W;n;h}(t_0,\ldots,t_h)$$

$$= \sum_{q=0}^{n} \sum_{\substack{\ell_0+\ldots+\ell_{h-1}=q \\ \ell_0,\ldots,\ell_{h-1}\geq 0}} N(\ell_0,\ldots,\ell_{h-1}, n-q)t_0^{\ell_0}\ldots t_{h-1}^{\ell_{h-1}}t_h^{n-q} .$$

If $q = 0$, then the inner sum is 0 since each of its summands is $N(0,\ldots,0,n)t_h^n = 0$ by (3.4). Hence we have

$$\Psi_{W;h}(z;t_0,\ldots,t_h) = \sum_{n=1}^{\infty} \frac{1}{n!} z^n \sum_{q=1}^{n} \sum_{\substack{\ell_0+\ldots+\ell_{h-1}=q \\ \ell_0,\ldots,\ell_{h-1}\geq 0}} N(\ell_0,\ldots,\ell_{h-1}, n-q)t_0^{\ell_0}\ldots t_{h-1}^{\ell_{h-1}}t_h^{n-q}$$

$$= \sum_{q=1}^{\infty} z^q \sum_{\substack{\ell_0+\ldots+\ell_{h-1}=q \\ \ell_0,\ldots,\ell_{h-1}\geq 0}} t_0^{\ell_0}\ldots t_{h-1}^{\ell_{h-1}} \sum_{n=q}^{\infty} \frac{1}{n!} N(\ell_0,\ldots,\ell_{h-1}, n-q)(zt_h)^{n-q} .$$

By (3.6) the innermost sum is

$$\sum_{n=q}^{\infty} \frac{1}{n!} \binom{n}{n-q} \ell_{h-1}^{n-q} N_{W;q}(\ell_0,\ldots,\ell_{h-1})(zt_h)^{n-q}$$

$$= \frac{1}{q!} N_{W;q}(\ell_0,\ldots,\ell_{h-1})\, e^{\ell_{h-1}zt_h} .$$

Consequently,

$$\Psi_{W;h}(z;t_0,\ldots,t_h)$$

$$= \sum_{q=1}^{\infty} \frac{1}{q!} z^q \sum_{\substack{\ell_0+\ldots+\ell_{h-1}=q \\ \ell_0,\ldots,\ell_{h-1}\geq 0}} N_{W;q}(\ell_0,\ldots,\ell_{h-1})$$

$$t_0^{\ell_0}\ldots t_{h-2}^{\ell_{h-2}}(t_{h-1}e^{zt_h})^{\ell_{h-1}} .$$

Then (3.15) implies (3.17).

Note that (3.15) and (3.2) give

$$(3.18)\qquad \Psi_{W;0}(z;t) = \sum_{n=1}^{\infty} \frac{1}{n!} z^n N_{W;n}(n)t^n$$

$$= \sum_{n=1}^{\infty} \frac{1}{n!} w_n (zt)^n = \Phi_W(zt)$$

by (1.1). If W has the cardinality property, then (3.17) yields

$$(3.19)\qquad \Psi_{W;1}(z;t_0,t_1) = \Psi_{W;0}(z;t_0 e^{zt_1}) = \Phi_W(zt_0 e^{zt_1}) .$$

Now (3.18) and (3.19) suggest that a general relation of this kind may hold for $\Psi_{W;h}(z;t_0,\ldots,t_h)$. In order to express this we require certain functions G_h of h+1 variables defined by:

$$(3.20)\qquad G_0(z_0) = z_0 , \qquad G_1(z_0,z_1) = z_0 e^{z_1} ,$$

$$(3.21)\qquad G_h(z_0,\ldots,z_h) = G_{h-1}(z_0,\ldots,z_{h-2},z_{h-1}e^{z_h})$$

if $h \geq 2$.

<u>Lemma 3</u>. If $h \geq 1$, then

$$(3.22)\qquad G_h(z_0,\ldots,z_h) = z_0 e^{G_{h-1}(z_1,\ldots,z_h)}$$

Proof. This is clear for $h = 1$. We therefore assume the result holds for $h-1$ where $h \geq 2$. Then (3.21) and the induction hypothesis give

$$\begin{aligned} G_h(z_0, \dots, z_h) &= G_{h-1}(z_0, \dots, z_{h-2}, z_{h-1}e^{z_h}) \\ &= z_0 \exp G_{h-2}(z_1, \dots, z_{h-2}, z_{h-1}e^{z_h}) \\ &= z_0 \exp G_{h-1}(z_1, \dots, z_{h-1}, z_h) \end{aligned}$$

which is just (3.22).

Corollary 1. If $h \geq 1$, then

$$G_h(z_0 t, z_1, \dots, z_h) = t\, G_h(z_0, z_1, \dots, z_h). \tag{3.23}$$

Proof. This is a consequence of (3.22).

Corollary 2. If $h \geq 0$ and $\max\{|z_0|, \dots, |z_h|\} \leq 1/4$, then $|G_h(z_0, \dots, z_h)| < 1/e$.

Proof. This is clear for $h = 0$. Hence, assume it holds for $h - 1$ where $h \geq 1$. Then

$$\begin{aligned} |G_h(z_0, \dots, z_h)| &\leq |z_0| \exp |G_{h-1}(z_1, \dots, z_h)| \\ &\leq \frac{1}{4} \exp\left(\frac{1}{e}\right) < \frac{1}{e}. \end{aligned}$$

In this result, we could have replaced $1/4$ by any number not exceeding $\exp(-1-e^{-1})$. We now derive a preliminary form of the composition theorem.

Lemma 4. Let $h \geq 0$, $\max\{|zt_0|, \dots, |zt_h|\} < 1/e$, and $|G_m(zt_{h-m}, \dots, zt_h)| < 1/e$ for $1 \leq m \leq h$. If W has the cardinality property, then

$$\Psi_{W;h}(z; t_0, \dots, t_h) = \Phi_W\left(G_h(zt_0, \dots, zt_h)\right). \tag{3.24}$$

Proof. For $h = 0$, this follows from (3.18). Now suppose that the result holds for $h - 1$ where $h \geq 1$. Then

$$|zt_{h-1}e^{zt_h}| = |G_1(zt_{h-1}, zt_h)| < 1/e$$

on taking $m = 1$ in our hypotheses. Consequently, Lemma 2 gives

$$(3.25) \qquad \Psi_{W;h}(z;t_0, \ldots, t_h) = \Psi_{W;h-1}(z;t_0, \ldots, t_{h-2}, t_{h-1}e^{zt_h}).$$

Moreover, if $1 \leq m \leq h-1$, then (3.21) gives

$$|G_m(zt_{h-1-m}, \ldots, zt_{h-2}, zt_{h-1}e^{zt_h})|$$

$$= |G_M(zt_{h-M}, \ldots, zt_{h-1}, zt_h)| < 1/e$$

as a result of our hypotheses and the fact that $M \equiv m + 1 \in [2, h]$. Consequently (3.25) and the induction hypothesis show that

$$\Psi_{W;h}(z;t_0, \ldots, t_h) = \Phi_W\left(G_{h-1}(zt_0, \ldots, zt_{h-2}, zt_{h-1}e^{zt_h})\right)$$

$$= \Phi_W\left(G_h(zt_0, \ldots, zt_{h-1}, zt_h)\right)$$

as a result of (3.21). This completes the induction.

Theorem 2 (Composition Theorem). Let $h \geq 0$ and $\max\{|zt_0|\ldots, |zt_h|\} \leq 1/4$. If W has the cardinality property, then (3.24) holds.

Proof. Corollary 2 of Lemma 3 shows that all the hypotheses of Lemma 4 are satisfied. Hence (3.24) follows.

This result clearly reveals how the generating function $\Psi_{W;h}$ depends on the prescribed matrix W. Namely, there is a universal function G_h, independent of W, such that the composition formula $\Psi_{W;h} = \Phi_W \circ G_h$ holds.

We now obtain some important specializations. Let $E_0(z, u) = zu$ and $E_h(z, u) = G_h(z, \ldots, z, zu)$ if $h \geq 1$. As a result of (3.22) we see that

$$(3.26) \qquad E_h(z, u) = z\,e^{G_{h-1}(z, \ldots, z, zu)} = z\,e^{E_{h-1}(z, u)}$$

if $h \geq 2$; and if $h \geq 1$, then (3.20) gives

$$E_1(z, u) = G_1(z, zu) = z\,e^{zu} = z\,e^{E_0(z, u)}$$

so that (3.26) holds for all $h \geq 1$. We also define $G_h(z) = G_h(z, \ldots, z)$ for all $h \geq 0$. Hence $G_h(z) = E_h(z, 1)$ so that

$$(3.27) \qquad G_0(z) = z, \quad G_h(z) = z\,e^{G_{h-1}(z)} \quad \text{if } h \geq 1.$$

The functions $G_h(z)$ were introduced by Rényi and Szekeres [16]. Further, we have $E_h(z, u/z) = F_h(z, u)$ where the $F_h(z, u)$ are the functions of Rényi and Szekeres satisfying $F_0(z, u) = u$ and $F_h(z, u) = z \exp\{F_{h-1}(z, u)\}$ for $h \geq 1$.

Corollary 1. Let $h \geq 1$ and $\max\{|z|, |zt|, |zu|\} \leq 1/4$. If W has the cardinality property then

$$(3.28) \qquad \sum_{n=1}^{\infty} \sum_{H=0}^{\infty} \sum_{\ell=0}^{\infty} Z_{W;n;h,H,\ell} \frac{z^n}{n!} u^H t^\ell = \Phi_W(t\,E_h(z, u))$$

so that $Z_{h,H,\ell}$ is the coefficient of $\frac{z^n}{n!} u^H t^\ell$ in the expansion of $\Phi_W(t\,E_h(z, u))$.

Proof. We set $t_0 = t$, $t_h = u$ and $t_1 = \ldots = t_{h-1} = 1$ in Theorem 2. On the one hand, (3.23) gives

$$\Psi_{W;h}(z;t, 1, \ldots, 1, u) = \Phi_W\big(G_h(zt, z, \ldots, z, zu)\big)$$

$$= \Phi_W(t\,G_h(z, z, \ldots, z, zu)) = \Phi_W\big(tE_h(z, u)\big)$$

if $h \geq 1$. On the other hand, (3.15) gives

$$\Psi_{W;h}(z;t,1,\ldots,1,u)$$

$$= \sum_{n=1}^{\infty} \frac{1}{n!} z^n \sum_{\substack{\ell_0+\ldots+\ell_h=n \\ \ell_0,\ldots,\ell_h\geq 0}} N(\ell_0,\ldots,\ell_h) t^{\ell_0} u^{\ell_h}$$

$$(3.29) \qquad = \sum_{n=1}^{\infty} \frac{1}{n!} z^n \sum_{\ell_h=0}^{n} u^{\ell_h} \sum_{\substack{\ell_0+\ldots+\ell_{h-1}=n-\ell_h \\ \ell_0,\ldots,\ell_{h-1}\geq 0}} N(\ell_0,\ldots,\ell_h) t^{\ell_0}$$

$$= \sum_{n=1}^{\infty} \frac{1}{n!} z^n \sum_{H=0}^{n} u^H \sum_{\ell_0=0}^{n-H} t^{\ell_0} \sum_{\substack{\ell_1+\ldots+\ell_{h-1}=n-\ell_0-H \\ \ell_1,\ldots,\ell_{h-1}\geq 0}} N(\ell_0,\ldots,\ell_{h-1},H)$$

$$= \sum_{n=1}^{\infty} \frac{1}{n!} z^n \sum_{H=0}^{n} u^H \sum_{\ell=0}^{n-H} t^{\ell} Z_{h,H,\ell}$$

by (3.5) provided $h \geq 2$. The last expression is just the left side of (3.28) by virtue of (2.4), (2.11) and (2.9). Thus the result holds for $h \geq 2$.

If $h = 1$, then (3.29) is still valid and its right side is

$$\sum_{n=1}^{\infty} \frac{1}{n!} z^n \sum_{\ell_1=0}^{n} u^{\ell_1} t^{n-\ell_1} N(n-\ell_1, \ell_1).$$

By (2.10) and (3.3)

$$\sum_{\ell_1=0}^{n} u^{\ell_1} \sum_{\ell_0=0}^{n-\ell_1} t^{\ell_0} Z_{1,\ell_1,\ell_0} = \sum_{\ell_1=0}^{n} u^{\ell_1} t^{n-\ell_1} Z_{1,\ell_1,n-\ell_1}$$

$$= \sum_{\ell_1=0}^{n} u^{\ell_1} t^{n-\ell_1} N(n-\ell_1, \ell_1) .$$

Thus, if $h = 1$ the right side of (3.29) is

$$\sum_{n=1}^{\infty} \frac{1}{n!} z^n \sum_{\ell_1=0}^{n} u^{\ell_1} \sum_{\ell_0=0}^{n-\ell_1} t^{\ell_0} Z_{1,\ell_1,\ell_0}$$

$$= \sum_{n=1}^{\infty} \frac{1}{n!} z^n \sum_{H=0}^{n} u^H \sum_{\ell=0}^{n-H} t^{\ell} Z_{1,H,\ell} .$$

On using (2.4), the last expression once more becomes the left side of (3.28).

Corollary 2. Let $h \geq 1$ and let W have the cardinality property. If $\max\{|z|, |zu|\} \leq 1/4$, then

$$(3.30) \qquad \sum_{n=1}^{\infty} \sum_{H=0}^{\infty} V_{W;n;h,H} \frac{z^n}{n!} u^H = \Phi_W(E_h(z, u))$$

so that $V_{h,H}$ is the coefficient of $\frac{z^n}{n!} u^H$ in the expansion of $\Phi_W(E_h(z, u))$. And if $\max\{|z|, |zt|\} \leq 1/4$, then

$$(3.31) \qquad \sum_{n=1}^{\infty} \sum_{\ell=0}^{\infty} Y_{W;n;h,\ell} \frac{z^n}{n!} t^{\ell} = \Phi_W(t\, G_h(z))$$

so that $Y_{h,\ell}$ is the coefficient of $\frac{z^n}{n!} t^{\ell}$ in the expansion of $\Phi_W(t\, G_h(z))$.

Proof. First, let $t = 1$ in (3.28); then (2.13), (2.4) and (2.6) yield (3.30). Next, take $u = 1$ in (3.28) and apply (2.18), (2.4) and (2 9) to get (3.31).

Corollary 3. If $h \geq 0$ and W has the cardinality property, then for $|z| \leq 1/4$.

$$\Psi_{W;h}(z) = \sum_{n=1}^{\infty} V_{W;n;h} \frac{z^n}{n!} = \Phi_W(G_h(z)) \tag{3.32}$$

so that V_h is the coefficient of $\frac{z^n}{n!}$ in the expansion of $\Phi_W(G_h(z))$.

Proof. If $h = 0$ this is a consequence of (2.26) and (1.1). If $h \geq 1$, we set $t = 1$ in (3.31) and apply (2.20) and (2.16)

We have now completed the proof of (1.3) and various generalizations of it.

4. Examples.

Before proceeding to these, it is helpful to determine the combinatorial significance of the universal functions $G_h(z)$. To this end, it is useful to define $\Psi_h(z) = \frac{1}{z} G_{h+1}(z)$ for $h \geq 0$. Then (3.27) yields

$$\Psi_0(z) = e^z, \quad \Psi_h(z) = e^{z\Psi_{h-1}(z)} \quad \text{if } h \geq 1, \tag{4.1}$$

and we put

$$\Psi_h(z) = \sum_{n=0}^{\infty} U_{hn} \frac{z^n}{n!} \quad \text{if } h \geq 0. \tag{4.2}$$

It is then clear from (45) of Riordan [20] (together with the corrected equation $S(y;0) = y$) that $\Psi_h(z)$ is the same as Riordan's $H(1, z; h)$ which enumerates the forests of rooted labeled trees of height not exceeding h; i.e., U_{hn} is the number of forests of rooted labeled trees with n vertices and height not exceeding h. Thus U_{hn} is the same as $H_n(1;h)$ which Riordan has tabulated for $h = 0, 1, \ldots, 7$ and $n = 1, 2, \ldots, 8$. For $h = 1$, a tabulation for $n = 1, 2, \ldots, 16$, and selected additional values, has been given by Harris and Schoenfeld [8]. We will return to this identification of U_{hn}

in Example 5.

Example 1. Let $W_{\ell r}$ be the set of all permutations on the ℓ letters in $X_{\ell r}$; thus $W_{\ell 1} = S_\ell$. Then W has the cardinality property and $w_\ell = \ell!$ for $\ell \geq 1$. Hence

$$\Phi_W(z) = \frac{z}{1-z} = \frac{1}{1-z} - 1 . \tag{4.3}$$

Now the condition (A') represents no restriction on α at all. Consequently V_h is just the number of functional digraphs on n vertices which have height at most h . Hence, also, Z is the number of functional digraphs with n vertices; so $Z = n^n$. We also see that Y_ℓ is just the number of functional digraphs on n letters which have exactly ℓ cyclical vertices. By (3.9), we get

$$Y_\ell = \binom{n}{\ell} \ell \cdot \ell! \, n^{n-\ell-1} = \frac{n!}{(n-\ell)!} \ell \, n^{n-\ell-1} \tag{4.4}$$

if $\ell \geq 1$; it holds also for $\ell = 0$. And (2.23) yields

$$n^n = Z = \sum_{\ell=1}^{n} \frac{n!}{(n-\ell)!} \ell \, n^{n-\ell-1} = n! \sum_{q=0}^{n-1} \frac{n^{q-1}}{q!} (n-q). \tag{4.5}$$

This identity, due to Abel, can also be obtained directly since the last summand is

$$\frac{n^q}{q!} - \frac{n^{q-1}}{(q-1)!} \qquad \text{if } q \geq 1 .$$

The result (4.4) is equivalent to (3.12) of Harris [6] and to (19) of Riordan [19]. Moreover, on using the tabulated values for U_{1n}, we find that

$$\Phi_W(G_2(z)) = \frac{1}{1-z\Psi_1(z)} - 1 = z + 4\frac{z^2}{2!} + 27\frac{z^3}{3!} + 232\frac{z^4}{4!} + \ldots$$

so that $V_{W;4;2} = 232$ by (3.32); thus the number of functional digraphs with 4 vertices and height at most 2 is 232.

Example 2. Let $W_{\ell r}$ be the set of all cyclical permutations of length ℓ on the ℓ letters in $X_{\ell r}$. Then W has the cardinality property and $w_\ell = (\ell - 1)!$ if $\ell \geq 1$.

Hence $\Phi_W(z) = -\log(1-z)$. If α satisfies (A') or (A), then the main permutation α^* is connected and hence α is as well. Consequently, V_h is the number of connected mappings on n letters which have height at most h. Likewise, Y_ℓ is the number of connected mappings on n letters having exactly ℓ cyclical vertices. Then (3.9) gives

$$(4.6) \qquad Y_\ell = \binom{n}{\ell}\ell\cdot(\ell-1)!\, n^{n-\ell-1} = \frac{n!}{(n-\ell)!} n^{n-\ell-1} \quad \text{if } \ell \geq 1 .$$

Consequently, (2.23) shows that Z, the number of connected mappings on n letters, is just

$$(4.7) \qquad Z = \sum_{\ell=1}^{n} \frac{n!}{(n-\ell)!} n^{n-\ell-1} = n! \sum_{q=0}^{n-1} \frac{n^{q-1}}{q!} .$$

This is contained in a theorem of Katz [13], in (3.31) of Harris [6], in (10) of Riordan [19] and (5) of Rényi [15]. We can also make the expansion

$$\Phi_W(G_2(z)) = -\log\{1-z\,\Psi_1(z)\}$$

$$= \frac{z}{1!} + 3\frac{z^2}{2!} + 17\frac{z^3}{3!} + 118\frac{z^4}{4!} + 1029\frac{z^5}{5!} + \ldots$$

and thereby show that $V_{W;5;2} = 1029$. That is, the number of connected functional digraphs with 5 vertices and height not exceeding 2 in 1029.

<u>Example 3.</u> Let $W_{\ell r}$ be the set of all <u>even</u> permutations on the ℓ letters in $X_{\ell r}$; thus $W_{\ell 1} = A_\ell$, the alternating group on ℓ letters. Then W has the cardinality property, $w_1 = 1$ and $w_\ell = \ell!/2$ if $\ell \geq 2$. Hence

$$(4.8) \qquad \Phi_W(z) = z + \sum_{\ell=2}^{\infty} \frac{1}{2}\ell!\,\frac{z^\ell}{\ell!} = z + \frac{z^2}{2(1-z)} .$$

Now V_h is the number of mappings on n letters having height at most h and whose main permutation is even. Also Y_ℓ is the number of mappings on n letters whose main permutation is an even permutation on ℓ letters. Then (3.9) gives

(4.9) $$Y_\ell = \binom{n}{\ell}\ell \cdot \frac{1}{2}\ell!\; n^{n-\ell-1} = \frac{n!}{(n-\ell)!}\cdot\frac{\ell}{2}\cdot n^{n-\ell-1} \quad \text{if } \ell \geq 2$$

and $Y_1 = n^{n-1}$. Then Z, the total number of mappings on n letters having a main permutation which is even, is

(4.10) $$Z = \sum_{\ell=1}^{n} Y_\ell = \frac{1}{2}n^{n-1} + \frac{1}{2}n!\sum_{\ell=1}^{n}\frac{n^{n-\ell-1}}{(n-\ell)!}\ell = \frac{1}{2}(n+1)n^{n-1}$$

as a result of (4.5).

Example 4. As a slight modification of the previous example, let $W_{1r} = \phi$ for all $r \geq 1$. And if $\ell \geq 2$, let $W_{\ell r}$ be the set of all even permutations on the ℓ letters in $X_{\ell r}$ if r is even; but for $\ell \geq 2$ and r odd, let $W_{\ell r}$ be the set of all odd permutations on the ℓ letters in $X_{\ell r}$. Then W has the cardinality property with $w_1 = 0$ and $w_\ell = \ell!/2$ for all $\ell \geq 2$. Note, however, that the cycle structure of $W_{\ell 1}$ and $W_{\ell 2}$ are different for $\ell \geq 2$. Consider for example $\ell = 3$ and suppose that we had defined

$$X_{31} = \{1, 2, 3\},\; X_{32} = \{1, 2, 4\},\; X_{33} = \{1, 3, 4\},\; X_{34} = \{2, 3, 4\}.$$

Then, on using the standard notation for permutations,

$$W_{31} = \{(12)(3), (13)(2), (23)(1)\}, \quad W_{32} = \{(1)(2)(4), (124), (142)\}$$

$$W_{33} = \{(13)(4), (14)(3), (34)(1)\}, \quad W_{34} = \{(2)(3)(4), (234), (243)\}.$$

Hence W_{31} and W_{33} have the same cycle structure, as do W_{32} and W_{34}; but W_{31} and W_{32} have different cycle structures. Now

$$\Phi_W(z) = \frac{z^2}{2(1-z)}.$$

Although $Y_1 = 0$, Y_ℓ is still given by (4.9) for $\ell \geq 2$. Also $Z = \frac{1}{2}(n-1)n^{n-1}$.

Example 5. If $r \geq 1$, we let W_{1r} consist of the identity map on X_{1r} and we take $W_{\ell r} = \phi$ for all $\ell \geq 2$. Then W has the cardinality property, $w_1 = 1$, $w_\ell = 0$ for

$\ell \geq 2$ and $\Phi_W(z) = z$. Now V_h is the number of $\alpha \in T_n$ of height not exceeding h which have exactly one cyclical vertex and this has order 1. Since a cycle of length 1, a loop, can be interpreted as the root of a tree, we see that $V_{W;n;h} = t_{hn}$ where t_{hn} is the number of rooted labeled trees with n vertices and height not exceeding h. Then (3.9) gives $Y_\ell = 0$ for $\ell \geq 2$ and

$$Y_1 = n^{n-1} \tag{4.11}$$

where Y_1 is the number of rooted labeled trees on n letters. Hence the number of labeled trees on n letters, whether rooted or not, is

$$Y_1/n = n^{n-2} . \tag{4.12}$$

The last result is due to Cayley, and (4.11) can be found, for example, on p. 128 of Riordan [17]. Moreover, the composition theorem gives

$$\sum_{n=1}^{\infty} t_{hn} \frac{z^n}{n!} = \Psi_{W;h}(z) = \Phi_W(G_h(z)) = G_h(z) \tag{4.13}$$

which is (1.5) of Rényi and Szekeres [16]; see, also, (20) of Riordan [18]. Now, for $h \geq 1$,

$$G_h(z) = z\,\Psi_{h-1}(z) = \sum_{m=0}^{\infty} U_{h-1,\,m} \frac{z^{m+1}}{m!} = \sum_{n=1}^{\infty} n\,U_{h-1,\,n-1} \frac{z^n}{n!}$$

so that

$$t_{hn} = n\,U_{h-1,\,n-1} \quad \text{if } h \geq 1 . \tag{4.14}$$

We can also obtain results on H_{hn}, the number of forests of rooted labeled trees on n letters with height not exceeding h. For if we take a single such tree and delete its root $x \in X_n$, then we obtain a forest of rooted labeled trees with n-1 vertices in $X - \{x\}$ and height not exceeding h-1. For a fixed x, the number of such forests is $H_{h-1,\,n-1}$. As there are n values for x, we have

$$t_{hn} = n\,H_{h-1,\,n-1} . \tag{4.15}$$

Comparing this with (4.14) we see that $H_{hn} = U_{hn}$. This confirms the remarks at the beginning of this section concerning the combinatorial significance of U_{hn}.

In (47) of Riordan [20], there can be found a formula (with a misprint) connecting the U_{hn} with the $U_{h-1,n}$. And in [11], we have given an asymptotic formula for U_{hn} for fixed h and large n.

Example 6. We modify the preceding example by defining $W_{\ell r}$ to consist of the identity map on $X_{\ell r}$ for every positive ℓ and r. Then W has the cardinality property, $w_\ell = 1$ and $\Phi_W(z) = e^z - 1$. We easily see that V_h is the number of functional digraphs $\alpha \in T_n$ whose heights do not exceed h and whose cycles are loops. Alternatively, V_h is the number of α in the semigroup T_n such that $\alpha^{h+1} = \alpha^h$.

Such α can be considered to be generalized idempotents of T_n - the customary idempotents, satisfying $\alpha^2 = \alpha$, correspond to $h = 1$. The case $h = 1$ has been dealt with previously by Harris and Schoenfeld [8, 9], Harris [7] and Tainiter [21]. In this case, the generating function for V_1 is, by (1.3),

$$\exp G_1(z) - 1 = \exp(z\, e^z) - 1$$

in agreement with previous work.

For general h, we can again interpret the loops of α as roots; we then see that V_h is the number of forests of rooted labeled trees with n vertices and height not exceeding h. Thus, $V_h = U_{hn}$ and this can also be verified by the composition formula. By (3.9),

$$Y_\ell = \binom{n}{\ell} \ell\, n^{n-\ell-1} = \binom{n-1}{\ell-1} n^{n-\ell} \quad \text{if } \ell \geq 1 \tag{4.16}$$

where Y_ℓ is now the number of forests with n vertices which have ℓ rooted labeled trees. And (2.23) gives

$$Z = \sum_{\ell=1}^{n} \binom{n-1}{\ell-1} n^{n-\ell} = \sum_{m=0}^{n-1} \binom{n-1}{m} n^{n-1-m} = (n+1)^{n-1} \tag{4.17}$$

as the number of forests of rooted labeled trees with n vertices; this has been given previously on p. 102 of Riordan [20].

Consider now a labeled tree with n vertices which has ℓ branches entering the root. If we delete its root, then we obtain a forest with n-1 vertices and ℓ rooted labeled trees. Analogously to (4.15), we therefore see that the number of labeled trees with n vertices and ℓ branches entering the root is, by (4.16),

$$(4.18) \qquad n\, Y_{W;n-1,\ell} = n \binom{n-2}{\ell-1}(n-1)^{n-1-\ell} \quad \text{if } \ell \geq 1,\ n \geq 2.$$

Hence the number of such trees which have a specified vertex joined to ℓ other vertices is

$$(4.19) \qquad Y_{W;n-1,\ell} = \binom{n-2}{\ell-1}(n-1)^{n-1-\ell} \quad \text{if } \ell \geq 1,\ n \geq 2\,.$$

This result has been given by Clarke [2]. From (4.18) and (4.17) we again derive (4.12) for the number of labeled trees with n vertices.

Example 7. Let d be a fixed positive integer and let $W_{\ell r}$ consist of all the permutations $\alpha \in T_n$ on the ℓ letters in $X_{\ell r}$ which are products of disjoint cycles each of length d. (If d = 1, this is the previous example.) Then W has the cardinality property. Here V_h is the number of functional digraphs on n letters which have height at most h and all of whose cycles have length d.

If $\alpha^* \in W_{\ell r}$ has j components, then $\ell = jd$. Hence $W_{\ell r} = \phi$ if $d \nmid \ell$; in this case, $w_\ell = 0$. But if $\ell = jd$, then the number of ways of assigning the jd cyclical vertices to j component sets each having d elements is just

$$\frac{(jd)!}{j!\ d!^{\,j}}\,.$$

From the d elements in each of the j component sets, we can form (d-1)! distinct cycles of length d. Hence

$$(4.20) \qquad w_{jd} = (d-1)!^{\,j} \cdot \frac{(jd)!}{j!\ d!^{\,j}} = \frac{(jd)!}{j!\ d^{j}} \quad \text{if } j \geq 1\,.$$

So

$$(4.21) \qquad \Phi_W(z) = \sum_{j=1}^{\infty} \frac{(jd)!}{j!\, d^j} \cdot \frac{z^{jd}}{(j\, d)!} = \sum_{j=1}^{\infty} \frac{1}{j!} \left(\frac{z^d}{d}\right)^j = \Lambda_d(z) - 1$$

where

$$(4.22) \qquad \Lambda_d(z) = \exp(z^d/d) .$$

Also Y_ℓ is the number of $\alpha \in T_n$ having ℓ cyclical vertices each of order d. Hence (3.9) shows that $Y_\ell = 0$ if $d \nmid \ell$ and for positive integral j

$$(4.23) \quad Y_{jd} = \binom{n}{jd} jd \cdot \frac{(jd)!}{j!\, d^j} \cdot n^{n-jd-1} = \frac{n!\, n^{n-jd-1}}{(n-jd)!\,(j-1)!\, d^{j-1}} .$$

Finally, the total number of functional digraphs on n letters having all of its cycles of length d is

$$(4.24) \qquad Z = \sum_{\ell=1}^{n} Y_\ell = \sum_{1 \le j \le n/d} Y_{jd} = n! \sum_{j=1}^{[n/d]} \frac{n^{n-jd-1}}{(n-jd)!\,(j-1)!\, d^{j-1}} .$$

For $d = 1$, these results generalize Example 6. In the special case $d = 2$ and $h = 1$, Corollary 3 of the composition theorem implies that the coefficient of $z^n/n!$ in the expansion of

$$\exp\{\tfrac{1}{2} G_1^2(z)\} - 1 = \exp\left(\tfrac{1}{2} z^2 e^{2z}\right) - 1$$

$$= \frac{z^2}{2!} + 6\frac{z^3}{3!} + 27\frac{z^4}{4!} + 140\frac{z^5}{5!} + \ldots$$

is $V_{W;n;1}$. The last quantity is the number of functional digraphs on n letters which have height at most 1 and which are such that all their cyclical elements have order 2. In particular, $V_{W;4;1} = 27$.

Example 8. We can generalize Example 6 in another way. Let k be a fixed positive integer and let $W_{\ell r}$ be the set of all permutations on the ℓ letters in $X_{\ell r}$ such that the length of every component cycle is a divisor of k. Of course, Example 6 corresponds to $k = 1$. Again, W has the cardinality property. It is clear that $w_\ell = w_{\ell 1}$ is just the number of elements $\beta \in S_\ell$ such that $\beta^k = \varepsilon_\ell$ where ε_ℓ is the identity element in the group S_ℓ. As a result, the work of Chowla, Herstein and Scott [1] shows that the generating function $\Phi_W(z)$ of w_ℓ is given by

$$(4.25) \qquad \Phi_W(z) = \exp\left(\sum_{d|k} z^d/d\right) - 1 = \prod_{d|k} \Lambda_d(z) - 1 .$$

This result may also be proved directly as follows. Let k have exactly r distinct divisors $1 = d_1 < d_2 < \ldots < d_r = k$. If $\ell \geq 1$ then w_ℓ is the number of permutations in S_ℓ each of whose cycles has one of the lengths $d_1, d_2, \ldots, d_r$. Hence, if $\ell \geq 1$ then

$$(4.26) \qquad w_\ell = \sum_{\substack{j_1, \ldots, j_r \geq 0 \\ j_1 d_1 + \ldots + j_r d_r = \ell}} \pi(j_1, \ldots, j_r)$$

where $\pi(j_1 \ldots, j_r)$ is the number of permutations in S_ℓ having j_1 cycles of length d_1, j_2 cycles of length d_2, $\ldots, j_r$ cycles of length d_r. To evaluate this, note that the number, ν, of ways of distributing the ℓ elements in X_ℓ into r disjoint sets $R_1, R_2, \ldots, R_r$ containing $j_1 d_1$, $j_2 d_2$, $\ldots, j_r d_r$ elements respectively is

$$\nu = \frac{\ell!}{(j_1 d_1)!\,(j_2 d_2)! \ldots (j_r d_r)!} .$$

Moreover, by (4.20) the number of ways λ_m of assigning the $j_m d_m$ elements of R_m into j_m cycles each of length d_m is

$$\lambda_m = \frac{(j_m d_m)!}{j_m!\, d_m^{\,j_m}} .$$

Hence

$$\pi(j_1, \ldots, j_r) = \nu\lambda_1 \ldots \lambda_r = \frac{\ell!}{j_1! \ldots j_r!\, d_1^{j_1} \ldots d_r^{j_r}} .$$

Consequently, (4.26) gives for $\ell \geq 1$

$$(4.27) \qquad w_\ell = \sum_{\substack{j_1, \ldots, j_r \geq 0 \\ j_1 d_1 + \ldots + j_r d_r = \ell}} \frac{\ell!}{j_1! \ldots j_r!\, d_1^{j_1} \ldots d_r^{j_r}} .$$

Therefore

$$\Phi_W(z) = \sum_{\ell=1}^{\infty} \frac{1}{\ell!} w_\ell z^\ell = \sum_{\ell=1}^{\infty} z^\ell \sum_{\substack{j_1, \ldots, j_r \geq 0 \\ j_1 d_1 + \ldots + j_r d_r = \ell}} \frac{1}{j_1! \ldots j_r!\, d_1^{j_1} \ldots d_r^{j_r}}$$

$$= \sum_{\ell=0}^{\infty} \sum_{\substack{j_1, \ldots, j_r \geq 0 \\ j_1 d_1 + \ldots + j_r d_r = \ell}} \frac{z^{j_1 d_1 + \ldots + j_r d_r}}{j_1! \ldots j_r!\, d_1^{j_1} \ldots d_r^{j_r}} - 1$$

$$= \sum_{j_1=0}^{\infty} \cdots \sum_{j_r=0}^{\infty} \frac{1}{j_1!}\left(\frac{z^{d_1}}{d_1}\right)^{j_1} \cdots \frac{1}{j_r!}\left(\frac{z^{d_r}}{d_r}\right)^{j_r} - 1$$

$$= \left(\exp \frac{z^{d_1}}{d_1}\right) \cdots \left(\exp \frac{z^{d_r}}{d_r}\right) - 1$$

which is just (4.25).

We easily see that V_h is the number of functional digraphs $\alpha \in T_n$ whose heights do not exceed h and which are such that for every cyclic vertex y of α the length of $C_\alpha(y)$ divides k. Expressed in terms of the semigroup T_n, the quantity V_h is the number of $\alpha \in T_n$ such that $\alpha^{h+k} = \alpha^h$. Such α may also be thought of as generalized idempotents.

If k is a prime p, then $r = 2$, $d_1 = 1$, $d_2 = p$ and (4.27) yields for $\ell \geq 1$

$$(4.28)\qquad w_\ell = \sum_{\substack{j,\, s \geq 0 \\ j + sp = \ell}} \frac{\ell!}{j!\; s!\; p^s} = \sum_{s=0}^{[\ell/p]} \frac{\ell!}{s!\,(\ell - ps)!\; p^s}.$$

This is due to Jacobsthal [12]. In [10], we gave the generating function $\Psi_{W;h}(z)$ of $V_{W;n;h}$ in terms of $\Psi_h(z)$; now it suffices to use the composition theorem. From (3.9), we see that

$$(4.29)\qquad Y_\ell = \binom{n}{\ell} \ell\, n^{n-\ell-1} \sum_{s=0}^{[\ell/p]} \frac{\ell!}{s!\,(\ell - ps)!\, p^s} \qquad \text{if } \ell \geq 1$$

where Y_ℓ is the number of $\alpha \in T_n$ such that α has ℓ cyclic vertices each of which has order 1 or p. For fixed p, Moser and Wyman [14] have derived an asymptotic formula for w_ℓ as $\ell \to \infty$.

Example 9. We conclude this section by considering those $\alpha \in T_n$ with no fixed points; that is, $\alpha(x) \neq x$ for every $x \in X_n$. These α are the functional digraphs with no cycles of length 1. Hence $W_{\ell r}$ consists of all permutations on the ℓ letters in $X_{\ell r}$ with the property that none of the ℓ letters is mapped into itself. Clearly, W has the cardinality property. Moreover, it is known (see, for example, Riordan [17, p. 59]) that $w_\ell = D_\ell$ where

$$D_\ell = \ell! \sum_{m=1}^{\ell} \frac{(-1)^m}{m!} = \ell!\,\{1 - 1 + \frac{1}{2!} - \frac{1}{3!} + \ldots + \frac{(-1)^\ell}{\ell!}\}$$

is the ℓ-th rencontres, displacement or derangement number. As

$$|D_\ell - \frac{\ell!}{e}| < \ell! \cdot \frac{1}{(\ell+1)!} = \frac{1}{\ell+1} \leq \frac{1}{2}$$

for $\ell \geq 1$, it follows that w_ℓ is the nearest integer to $\ell!/e$. Further, it is easily seen that

$$\Phi_W(z) = \sum_{\ell=1}^{\infty} \frac{D_\ell}{\ell!} z^\ell = \frac{e^{-z}}{1-z} - 1. \tag{4.30}$$

By (3.9), the number of $\alpha \in T_n$ having ℓ cyclical vertices none of which is of order 1 is

$$Y_\ell = \binom{n}{\ell} \ell D_\ell n^{n-\ell-1} = \binom{n-1}{\ell-1} n^{n-\ell} D_\ell \qquad \text{if } \ell \geq 1. \tag{4.31}$$

As the total number of mappings $\alpha \in T_n$ having no fixed points is just $(n-1)^n$, we get the identity

$$(n-1)^n = Z = \sum_{\ell=1}^{n} \binom{n-1}{\ell-1} n^{n-\ell} D_\ell \tag{4.32}$$

noted by Riordan on p. 184 of [19].

REFERENCES

1. S. Chowla, I. N. Herstein, and W. R. Scott, The solutions of $x^d = 1$ in symmetric groups, Norske Vid. Selsk. Forhandlinger, Trondheim, 25 (1952), 29-31.

2. L. E. Clarke, On Cayley's formula for counting trees, J. London Math. Soc. 33 (1958), 471-474.

3. J. Dénes, On transformations, transformation-semi-groups and graphs, in Theory of Graphs, Proceedings of the Colloquium held at Tihany, Hungary, September 1966, 65-75. Academic Press, New York, 1968.

4. J. Dénes, On some properties of commutator subsemi-groups, Publicationes Mathematicae (Debrecen), 15 (1968), 283-285.

5. Jósef Dénes, On graph representation of semigroups, Proceedings of the Calgary International Conference on Combinatorial Structures and their Applications, pp. 55-57. Edited by Richard Guy et al. Gordon and Breach Science Publishers, New York, London, Paris, 1970.

6. Bernard Harris, Probability distributions related to random mappings, Ann. Math. Statist. 31 (1960), 1045-1062.

7. Bernard Harris, A note on the number of idempotent elements in symmetric semigroups, Amer. Math. Monthly, 74 (1967), 1234-1235.

8. Bernard Harris and Lowell Schoenfeld, The number of idempotent elements in symmetric semigroups, J. Combinatorial Theory, 3 (1967), 122-135.

9. Bernard Harris and Lowell Schoenfeld, Asymptotic expansions for the coefficients of analytic functions, Illinois J. of Math. 12 (1968), 264-277.

10. Bernard Harris and Lowell Schoenfeld, The number of solutions of $\alpha^{j+p} = \alpha^j$ in symmetric semigroups, Proceedings of the Calgary International Conference on Combinatorial Structures and their Applications, pp. 155-157. Edited by Richard Guy et al. Gordon and Breach Science Publishers, New York, London, Paris, 1970.

11. Bernard Harris and Lowell Schoenfeld, The number of generalized idempotent elements in symmetric semigroups, MRC Technical Summary Report #998, 1969.

12. E. Jacobsthal, Sur le nombre d'éléments du groupe symétrique S_n dont l'ordre est un nombre premier. Norske Vid. Selsk. Forhandlinger, Trondheim, 21 (1949), 49-51.

13. Leo Katz, Probability of indecomposability of a random mapping function, Ann. Math. Statist., 26 (1955), 512-517.

14. Leo Moser and Max Wyman, On solutions of $x^d = 1$ in symmetric groups, Canadian J. Math. 7 (1955), 159-168.

15. A. Rényi, On connected graphs, I., Magyar Tud. Akad. Mat. Kutató Int. Közl., 4 (1959), 385-388.

16. A. Rényi and G. Szekeres, On the height of trees, J. Australian Math. Soc. 7 (1967), 497-507 .

17. John Riordan, An Introduction to Combinatorial Analysis, John Wiley and Sons, Inc., New York, 1958 .

18. John Riordan, The enumeration of trees by height and diameter, IBM Journal of Research and Development, 4 (1960), 473-478 .

19. John Riordan, Enumeration of linear graphs for mappings of finite sets, Ann. Math. Statist. 33 (1962), 178-185.

20. John Riordan, Forests of labeled trees, J. Combinatorial Theory, 5 (1968), 90-103.

21. M. Tainiter, A characterization of idempotents in semigroups, J. Combinatorial Theory, 5 (1968), 370-373.

Symbols for the Harris-Schoenfeld Paper

Index